Advances in Intelligent Systems and Computing

Volume 222

T0138180

Series Editor

J. Kacprzyk, Warsaw, Poland

For further volumes:
http://www.springer.com/series/11156

Mohd Saberi Mohamad · Loris Nanni
Miguel P. Rocha · Florentino Fdez-Riverola
Editors

7th International Conference on Practical Applications of Computational Biology and Bioinformatics

Editors
Mohd Saberi Mohamad
Faculty of Computing
Universiti Teknologi Malaysia
Johor
Malaysia

Loris Nanni
Dipartimento di Ingegneria dell'Informazione
University of Padua
Padua
Italy

Miguel P. Rocha
Departamento de Informática / CCTC
Universidade do Minho
Braga
Portugal

Florentino Fdez-Riverola
Department of Informatics
ESEI: Escuela Superior de Ingeniería
Informática
University of Vigo
Edificio Politécnico. Campus Universitario As
Lagoas s/n
Ourense
Spain

ISSN 2194-5357   ISSN 2194-5365   (electronic)
ISBN 978-3-319-00577-5   ISBN 978-3-319-00578-2   (eBook)
DOI 10.1007/978-3-319-00578-2
Springer Cham Heidelberg New York Dordrecht London

Library of Congress Control Number: 2013937504

Printed on acid-free paper

Springer is part of Springer Science+Business Media (www.springer.com)

Preface

The growth of the Bioinformatics and Computational Biology fields in recent years has been remarkable and its pace has not slowed down. The impressive capabilities of next generation sequencing technologies, together with novel and ever evolving distinct types of omics data technologies, have put an increasingly complex set of challenges for those fields. To address the problems posed by these huge amounts of data, and their inherent complexity, and also to address the multiple related tasks, for instance in biological modeling, there is the need to, more than ever, create multidisciplinary networks of collaborators, spanning computer scientists, mathematicians, biologists, doctors and many others.

The International Conference on Practical Applications of Computational Biology & Bioinformatics (PACBB) is an annual international meeting dedicated to emerging and challenging applied research in Bioinformatics and Computational Biology. Building on the success of previous events, the 7th edition of PACBB Conference will be held on 22-24 May 2013 in the University of Salamanca, Spain. In this occasion, a special issue published by the Journal of Integrative Bioinformatics will cover extended versions of selected articles.

This volume gathers the accepted contributions for the 7th edition of the PACBB Conference after being reviewed by different reviewers, from an international committee composed of 69 members from 9 countries. PACBB'13 technical program includes 19 papers spanning many different sub-fields in Bioinformatics and Computational Biology.

Therefore, this event will strongly promote the interaction of researchers from diverse fields and distinct research groups. The scientific content will definitely be challenging and will promote the improvement of the valuable work that is being carried out by the participants. Also, it will promote the education of young scientists, in a post-graduate level, in an interdisciplinary field.

We would like to thank all the contributing authors and sponsors: (Telefónica Digital, Indra, Ingeniería de Software Avanzado S.A, IBM, JCyL, IEEE Systems Man and Cybernetics Society Spain, AEPIA Asociación Española para la Inteligencia Artificial, APPIA Associação Portuguesa Para a Inteligência Artificial, CNRS Centre national de la recherche scientifique) as well as the members of the Program Committee and the Organizing Committee for their hard and highly valuable work and support. Their effort has helped to contribute to the success of the PACBB'13 event. PACBB'13 wouldn't exist without your assistance. This symposium is organized by the Bioinformatics, Intelligent System and Educational Technology Research Group (http://bisite.usal.es/) of the University of Salamanca and the Next Generation Computer System Group (http://sing.ei.uvigo.es/) of the University of Vigo.

Mohd Saberi Mohamad
Loris Nanni
PACBB'13 Programme Co-chairs

Miguel P. Rocha
Florentino Fdez-Riverola
PACBB'13 Organizing Co-chairs

Organization

General Co-chairs

Mohd Saberi Mohamad   Universiti Teknologi Malaysia, Malaysia
Loris Nanni   University of Padua, Italy
Miguel P. Rocha   CCTC, Univ. Minho, Portugal
Florentino Fdez-Riverola   University of Vigo, Spain

Program Committee

Alfonso Valencia   Structural Biology and BioComputing
    Programme (CNIO), Spain
Alicia Troncoso   University Pablo de Olavide, Spain
Amparo Alonso   University of A Coruña, Spain
Anália Lourenço   IBB/CEB, University of Minho, Portugal
Arlo Randall   University of California Irvine, USA
Armando Pinho   University of Aveiro, Portugal
Caludine Chaouiya   Gulbenkian Institute, Portugal
Christopher Henry   Argonne National Labs, USA
Daniel Glez-Peña   University of Vigo, Spain
David Posada   University of Vigo, Spain
Eva Lorenzo   University of Vigo, Spain
Fernando Diaz-Gómez   University of Valladolid, Spain
Florencio Pazos   CNB/CSIC, Madrid, Spain
Frank Klawonn   Ostafilia University of Applied Sciences,
    Wolfenbuettel, Germany
Giovani Librelotto   Universidade Federal de Santa Maria, Brasil
Gonzalo Gómez-López   UBio/CNIO, Spanish National Cancer
    Research Centre, Spain
Hagit Shatkay   University of Delaware, USA
Heri Ramampiaro   Norwegian University of Science and
    Technology, Trondheim, Norway

Isabel C. Rocha	IBB/CEB, University of Minho, Portugal
Jorge Vieira	IBMC, Porto, Portugal
José Luis Oliveira	University of Aveiro, Portugal
Juan Antonio García Ranea	University of Malaga, Spain
Juan F. de Paz	University of Salamanca, Spain
Julio R. Banga	IIM/CSIC, Vigo, Spain
Lourdes Borrajo	University of Vigo, Spain
Luis Figueiredo	European Bioinformatics Institute, EMBL-EBI, UK
Luis M. Rocha	Indiana University, USA
Manuel J. Maña López	University of Huelva, Spain
Manuel Rodriguez	University of Salamanca, Spain
Mª Araceli Sanchís de Miguel	University Carlos III of Madrid, Spain
Miguel Reboiro	University of Vigo, Spain
Nicholas Luscombe	European Bioinformatics Institute, EMBL-EBI
Nuno Fonseca	CRACS/INESC, Porto, Portugal
Reyes Pavón	University of Vigo, Spain
Rita Ascenso	Polytecnic Institute of Leiria, Portugal
Rosalía Laza	University of Vigo, Spain
Rui Brito	University of Coimbra, Portugal
Rui C. Mendes	CCTC, University of Minho, Portugal
Rui Camacho	LIAAD/FEUP, University of Porto, Portugal
Rui Rijo	Polytecnic Institute of Leiria, Portugal
Sara C. Madeira	IST/INESC ID, Lisbon, Portugal
Sérgio Deusdado	Polytecnic Institute of Bragança, Portugal
Thierry Lecroq	Univeristy of Rouen, France

Organising Committee

Juan M. Corchado	University of Salamanca, Spain
Javier Bajo	Pontifical University of Salamanca, Spain
Juan F. De Paz	University of Salamanca, Spain
Sara Rodríguez	University of Salamanca, Spain
Dante I. Tapia	University of Salamanca, Spain
Fernando de la Prieta Pintado	University of Salamanca, Spain
Davinia Carolina Zato Domínguez	University of Salamanca, Spain
Gabriel Villarrubia González	University of Salamanca, Spain
Alejandro Sánchez Yuste	University of Salamanca, Spain
Antonio Juan Sánchez Martín	University of Salamanca, Spain

Cristian I. Pinzón	University of Salamanca, Spain
Rosa Cano	University of Salamanca, Spain
Emilio S. Corchado	University of Salamanca, Spain
Eugenio Aguirre	University of Granada, Spain
Manuel P. Rubio	University of Salamanca, Spain
Belén Pérez Lancho	University of Salamanca, Spain
Angélica González Arrieta	University of Salamanca, Spain
Vivian F. López	University of Salamanca, Spain
Ana de Luís	University of Salamanca, Spain
Ana B. Gil	University of Salamanca, Spain
Mª Dolores Muñoz Vicente	University of Salamanca, Spain
Jesús García Herrero	University Carlos III of Madrid, Spain

Contents

Gene Functional Prediction Using Clustering Methods for the Analysis of Tomato Microarray Data

Liliana López-Kleine[1], José Romeo[2], and Francisco Torres-Avilés[2]

[1] Departamento de Estadística, Universidad Nacional de Colombia- Sede Bogotá, Colombia
llopezk@unal.edu.co
[2] Departamento de Matemática y Ciencia de la Computación, Universidad de Santiago de Chile, Chile
{jose.romeo,francisco.torres}@usach.cl

Abstract. Molecular mechanisms of plant-pathogen interaction have been studied thoroughly because of its importance for crop production and food supply. This knowledge is a starting point in order to identify new and specific resistance genes by detecting similar expression patterns. Here we evaluate the usefulness of clustering and data-mining methods to group together known plant resistance genes based on expression profiles. We conduct clustering separately on *P.infestans* inoculated and not-inoculated tomatoes and conclude that conducting the analysis separately is important for each condition, because grouping is different reflecting a characteristic behavior of resistance genes in presence of the pathogen.

Keywords: Microarray data analysis, clustering methods, functional gene prediction, resistance genes, plant immunity.

1 Introduction

Recent advances in genomic and post-genomic technologies have provided the opportunity to analyze genomic data publicly available in databases. Several molecular mechanisms can be understood better through the analysis of genomic data such as gene expression data. One of the molecular mechanisms, that needs to be better understood in order to reduce losses caused by plant pathogens is plant immunity. Losses caused by plant pathogens represent one of the most important limitations in crop production and therefore food supply. Understanding immunity responses in plants could help scoping with plant pathogens and even develop more resistant varieties.

Plant immunity depends on the recognition of conserved Microbial Associated Molecular Patterns (MAMPs) or strain-specific effectors by Pattern Recognition Receptors (PPRs) or resistance (R) proteins, triggering MTI (MAMP Triggered-Immunity) and ETI (Effector Triggered-Immunity), respectively. Upon recognition plants activate a complex network of responses that includes signal transduction pathways, novel protein interactions and coordinated changes in gene expression. Detailed information concerning specific and punctual interactions between effector

M.S. Mohamad et al. (Eds.): *7th International Conference on PACBB*, AISC 222, pp. 1–6.
DOI: 10.1007/978-3-319-00578-2_1 © Springer International Publishing Switzerland 2013

and resistance proteins has been accumulated in the last years, and in some cases, a global picture for some of these interactions has been established [1], [2]. Networks of immunity genes have been constructed for model plants such as Arabidopsis and rice primarily using yeast-two hybrid experiments [3], [4]. Nevertheless, each plant-pathogen interaction has its specificities and can conduct to the activation of immunity and resistance genes that were previously unknown. Moreover, functional relationships of genes that are not known to be resistance genes could be detected through the comparison of expression profiles during a specific plant disease. One of the most important genomic data used to reveal similarity patterns based on gene expression data and changes in gene expression profiles due to presence of pathogens, are microarray data. Microarray data are used in two ways: i) for detection of differentially expressed genes (comparing two conditions) and ii) for construction of clusters of genes with similar gene expression profile for functional gene prediction.

Here, we address the second objective and evaluate the usefulness of multivariate and data mining methods in order to cluster genes with similar expression patterns in a tomato time course experiment of pathogen infected tomatoes. The aim of applying this kind of analysis is grouping known resistance genes together with unknown resistance genes to achieve the functional prediction of new resistance genes or at least the functional participation in immunity processes of genes that are not known to participate in them.

In a previous work we defined a set of virulence factors based on literature [5] and were able to predict potential novel virulence factors based on linear and non-linear clustering methods combining all microarray experiments. Additionally, we applied the GEE modeling and influential analysis to these data [6] separately on both conditions and were able to confirm some virulence factors. Here, clustering and data mining methods were applied to microarray data available at the Tomato Expression Database (TED) at http://ted.bti.cornell.edu. These datasets compare two conditions in the field: healthy cherry tomato and *Phytophtora infestans* inoculated cherry tomato at 4 time points. Predictions were performed separately for each condition.

Conducing clustering and classification separately allowed us to determine strong differences in prediction indicating that the shift in gene expression of genes participating in immunity processes is strong. Moreover, we were able to cluster together more resistance genes when conducting the analysis on inoculated tomato, which indicates that their activation is needed to detect similar co-expression patterns and that information on most of these genes contained in healthy tomato, is not enough for clustering. We conclude that the here presented methodology, especially SVM, can be used for the prediction of new resistance genes, namely those that will be clustered together with genes that are not known to be resistance genes.

2   Methodology

2.1   Microarray Data

The data sets were obtained from TED. In this study we used experiments that were carried out using the TOM1 DNA chip platform and that are available at (http://ted.bti.cornell.edu/cgi-bin/TFGD/miame/experiment.cgi?ID=E022). We focused on the experiments carried out by Christine Smart and collaborators (Accession number E022) where gene expression profiling of infection of tomato by *P. infestans* in the field was studied. The goal of this experiment was to gain insight into the molecular basis of the compatible interaction between *P. infestans* and its hosts.

For this comparison four time points were available at 0, 12, 36 and 60 hours with 8 replicates of each condition (32 experiments) for 13440 tomato genes. We used the data from that experiment separately for inoculated plants (condition I) vs. non inoculated plants (condition NI).

Canonical immune protein domains (WRKY, TIR, NBS, kinase and LysM) from Pfam (http://pfam.sanger.ac.uk/) were searched in the tomato genome annotation file available at TED. We detected 174 genes coding for proteins with these domains and considered them as a set of "known resistance genes".

2.2   Classification Methods

We applied several unsupervised classification methods: k-means cluster analysis using the centroids obtained from a hierarchical cluster analysis [7], [8], Agglomerative Nesting (AGNES) [9], Divisive Analysis clustering - DIANA [9] in order to cluster genes in several groups. For all these methods we used two different distance measures: Euclidean and Manhattan or Taxicab metric [10].

We also applied Kohonen self organizing maps [11], which is a highly appreciated algorithm for its ability to classify data into two dimensions and is based on neural networks.

To define the best number of clusters, a visual inspection was performed using the dendrogram obtained from the hierarchical clustering method using both distance measures. Despite the total number of clusters, we were interested in obtaining a main group of known resistance genes.

Finally, we performed predictions of "resistance" based on a non-linear supervised classification kernel method called support vector machine (SVM) [12]. This method is a supervised method for which part of the data needs to be used for training the classifier. In order to train the SVM classifier a sample of 1/3 of the genes (4480) was used. Predictions were therefore obtained only for 2/3 of the genes (8960). In order to apply this method, microarray data need to be mapped in to a feature space by constructing a kernel. Here, we constructed a Gaussian kernel and tuned the sole parameter of this kernel (sigma) by leave-one-out crossvalidation on the training set optimizing the correct classification of known resistance genes into a homogeneous group. All data analyses were done in R-gui [13].

2.3   Validation of the Method

For all unsupervised clustering methods, we identified the two clusters that grouped most of the 174 known resistance genes and considered them to represent resistance gene clusters (RGC). Then, we identified the method that more accurately predicted resistance (allowed obtaining the most homogeneous RGCs) and used it as a reference to compare performance of the other methods. Furthermore, we counted resistance and non-resistance classifications in each RGC.

3   Results and Discussion

Here we focused on the ability of predicting known resistance genes in cherry tomato exposed to *P. infestans* in order to test different clustering and classification methods. One new proposal is that we classified separately for each condition based on the hypothesis that gene expression of inoculated tomato should reflect better the behavior of resistance genes involved in immunity processes against the pathogen than healthy tomato because less genes implicated in immunity will be active. Moreover, conducting clustering and classification separately considers the fact that gene expression of resistance genes in non-inoculated tomatoes reflects a basal expression and not necessarily a behavior due to the presence of the pathogen. Therefore, merging measures from both conditions could cause prediction errors associated to uncontrolled non-biological factors.

Classification results were different for each of the tested methods. The cluster number was chosen to be 55 for all clustering methods based on a visual inspection of a dendrogram. Differences between the two distances (Euclidean and Manhattan) were small, which let us conclude that each one can be used. All methods coincide in grouping together in the same RGC cluster 19/174 known resistance genes for not-inoculated tomato measures and 48/174 for inoculated ones. Most of the methods (K-means, AGNES and Kohonen) grouped resistance genes into two main clusters (Table 1). DIANA method grouped most known resistance genes in one cluster, but this method identified a very huge cluster with almost all genes (over 11000), therefore it could be discarded.

Table 1. Number of genes in the RGC obtained for each method (E: Euclidean distance, M: Manhattan distance). *2/3 of the dataset not including training data.

Method	Non-inoculated		Inoculated	
	Non-resistance	Resistance	Non-resistance	Resistance
K-means (E)	4818	58	4681	73
K-means (M)	4760	57	4681	73
AGNES (E)	6463	87	6863	99
AGNES (M)	6598	86	6756	99
DIANA (E)	11128	145	12632	158
DIANA (M)	4760	57	12325	156
Kohonen	5120	105	4323	75
SVM*	98	69	0	105

For the SVM classification no clusters were constructed, but resistance and non-resistance were considered as response variable for training a non-linear classifier. This classification method turned out to be the more accurate one because classification of non-resistance genes into the RGC is very small (i.e. the RGC is very homogeneous). SVM was used as the reference method (Table 1). The fact that SVM provides the most accurate classification could be due to the presence of non-linear patterns in gene expression that cannot be detected by other traditional methods. Nevertheless, some genes predicted to be resistance genes only by SVM could be due to noise and high variability of microarray data. Therefore, we advise using the predictions of all methods (discarding DIANA).

Some genes were identified as resistance genes in both conditions (Table 2). This could indicate that these genes have a basal expression and participate coordinately in house-keeping processes even when the plant is not exposed to a pathogen. Among known resistance genes these are the most suitable for pleiotropic functions that could not be related to plant immunity processes.

Table 2. List and function of common genes detected by the clustering and SVM algorithms on inoculated and non-inoculated tomato microarray data

Gene ID	Function
1-1-2.1.13.17	Avr9/Cf-9 rapidly elicited protein 146 [Nicotiana tabacum]
1-1-2.4.18.3	AvrPto-dependent Pto-interacting protein 3 [Lycopersicon esculentum]
1-1-3.3.11.4	WIZZ [Nicotiana tabacum]
1-1-3.4.10.21	Protein kinase-coding resistance protein [Nicotiana repanda]
1-1-4.1.17.2	Putative disease resistance protein RGA4, identical [Solanum bulbocastanum]
1-1-4.3.20.3	Disease resistance protein RGA2, putative [Ricinus communis]
1-1-7.4.19.21	WRKY transcription factor 26 [(Populus tomentosa x P. bolleana) x P. tomentosa]
1-1-8.2.15.5	Disease resistance protein RPS5, putative [Ricinus communis]
1-1-8.2.16.9	Avr9/Cf-9 induced kinase 1 [Nicotiana tabacum]
1-1-8.4.11.17	WRKY [Solanum lycopersicum]

The here obtained results suggest that conducting a separate analysis of both conditions is crucial. A shift in gene expression is detected in *P. infestans* inoculated cherry tomato indicating that most of the known resistance genes get activated in this condition.

Taking into account the present results, these clustering methodologies can be applied for the functional prediction of resistance genes through the selection of clusters with the highest frequency of known resistance genes. Moreover, other clustering methods could be tested, which could improve even more reliability of predictions. Prediction of new resistance genes should considered as those genes grouped together in the inoculated condition and not in the healthy tomatoes.

Moreover, this is a general approach that can be applied to different organisms for which functional gene prediction of a certain function of interest wants to be done and microarray datasets in which two or more conditions are available.

References

1. Pop, A., Huttenhower, C., Iyer-Pascussi, A., Benfey, P.N., Troyanskaya, O.G.: Integrated functional networks of process, tissue, and developmental stage specific interactions in Arabidopsis thaliana. BMC Syst. Biol. 4, 180 (2010)
2. Pritchard, L., Birch, P.: A systems biology perspective on plant-microbe interactions: Biochemical and structural targets of pathogen effectors. Plant Sci. 180, 584–603 (2011)
3. Mukhtar, S., Carvunis, A., Dreze, M., et al.: Independently Evolved Virulence Effectors Converge onto Hubs in a Plant Immune System Network. Science 333, 596–601 (2011)
4. Lee, H., Chah, O.K., Sheen, J.: Stem-cell-triggered immunity through CLV3p-FLS2 signalling. Nature 473, 376–379 (2011)
5. Lopez-Kleine, L., Torres-Avilés, F., Tejedor, F., Gordillo, L.A.: Virulence factor prediction in Streptococcus pyogenes using classification and clustering based on microarray data. Appl. Microbiol. Biotechnol. 93, 2091–2098 (2012)
6. Romeo, J., Torres-Avilés, F., Lopez-Kleine, L.: Influence analysis in Streptococcus pyogenes through Quasi Likelihood Model. Submitted to Mol. Genet. Genomics (2012)
7. Hartigan, J.A., Wong, M.A.: A k-means clustering algorithm. Appl. Stat. 28, 100–108 (1979)
8. Leiva-Valdebenito, S., Torres-Avilés, F.: Una revisión de los algoritmos de partición más comunes en el análisis de conglomerados: un estudio comparativo. Rev. Colomb. Estad. 33, 321–339 (2010)
9. Kaufman, L., Rousseeuw, P.J.: Finding Groups in Data: An Introduction to Cluster Analysis. Wiley, New York (1990)
10. Krause, E.F.: Taxicab Geometry: An Adventure in Non-Euclidean Geometry. Dover, New York (1986)
11. Kohonen, T.: Self-Organizing Maps. Springer, Heidelberg (2000)
12. Schölkopf, B., Smola, A.: Learning with Kernels: Support Vector Machi0nes, Regularization, Optimization, and Beyond. The MIT Press, Cambridge (2002)
13. R Development Core Team. R: A language and environment for statistical computing. R Foundation for Statistical Computing, Vienna, Austria (2012) ISBN 3-900051-07-0, http://www.R-project.org/

Analysis of Word Symmetries in Human Genomes Using Next-Generation Sequencing Data

Vera Afreixo[2], João M.O.S. Rodrigues[1], and Sara P. Garcia[1]

[1] Signal Processing Lab, IEETA and Department of Electronics Telecommunications and Informatics, University of Aveiro, 3810-193 Aveiro, Portugal
{spgarcia,jmr}@ua.pt
[2] CIDMA - Center for Research and Development in Mathematics and Applications, Department of Mathematics, University of Aveiro, 3810-193 Aveiro, Portugal
vera@ua.pt

Abstract. We investigate Chargaff's second parity rule and its extensions in the human genome, and evaluate its statistical significance. This phenomenon has been previously investigated in the reference human genome, but this sequence does not represent a proper sampling of the human population. With the 1000 genomes project, we have data from next-generation sequencing of different human individuals, constituting a sample of 1092 individuals. We explore and analyze this new type of data to evaluate the phenomenon of symmetry globally and for pairs of symmetric words.

Our methodology is based on measurements, traditional statistical tests and equivalence statistical tests using different parameters (e.g. mean, correlation coefficient).

We find that the global symmetries phenomenon is significant for word lengths smaller than 8. However, even when the global symmetry is significant, some symmetric word pairs do not present a significant positive correlation but a small or non positive correlation.

1 Introduction

Chargaff's second parity rule asserts that the percentage of complementary nucleotides should be similar in each of the two strands of a DNA sequence [11] [5] [12]. Different authors suggest and describe that there are similarities between the frequencies of words and of their inverted complements (which we call symmetry phenomenon), even for longer word lengths (e.g [10] [4] [3] [7] [14]). No previous work used genomes of several individuals from the same species to characterize the significance of the symmetry phenomenon in the species. In this work, we explore and characterize the significance of the symmetry phenomenon in the human genome, using data from multiple genomes made available by the 1,000 Genomes Project [2].

The contribution of this work is to present novel methodologies to explore the similarities between symmetric words using sequencing data obtained with next-generation methodologies. We explore the symmetry phenomenon in word lengths between 1 and 12 nucleotides.

M.S. Mohamad et al. (Eds.): *7th International Conference on PACBB*, AISC 222, pp. 7–13.
DOI: 10.1007/978-3-319-00578-2_2 © Springer International Publishing Switzerland 2013

2   Methods

We evaluate the symmetry phenomenon using word frequency counts. Words are interchangeably called k-mers. We study word lengths $k \in \{1, 2, ..., 12\}$. Our sample has $n = 1092$ human genomes. For each individual, all words of lengths k were counted, and for each word length, the word (w) and its corresponding symmetric word (w') counts are paired to obtain symmetric pair counts $(N_w, N_{w'})$.

Note that, the number of distinct k-mers is 4^k. For $i \in \{1, 2, ..., n\}$, N_w^i is the number of times the word w appears in the genome sequence of individual i and

$$\sum_w N_w^i = \sum_{w'} N_{w'}^i = S^i.$$

The corresponding relative frequencies are represented by $f_w^i = N_w^i / S^i$ and $f_{w'}^i = N_{w'}^i / S^i$.

2.1   Statistical Hypothesis Testing

Traditional statistical hypothesis testing may be used to assess differences. However, it is well known that when traditional hypothesis tests are applied to large data sets, any small effect is always deemed significant [9] [6] [8]. Furthermore, we want to evaluate if there are similarities, not differences, between the occurrence of symmetric words. To overcome this drawback, we use equivalence tests for accepting/rejecting the equivalence between symmetric words.

We studied the equivalence between pairs of symmetric words (w, w') using the ratio of the frequency of the symmetric pair R_w and a practical tolerance δ (> 1), and concluding the equivalence when $1/\delta < R_w < \delta$. Let μ_{R_w} denote the (population) mean of the ratio of the w word frequency and its corresponding reversed complement word frequency (ratio of the frequency of the symmetric pair). Let $\overline{R_w}$ denote the corresponding sample mean and for each individual i the ratio is given by $R_w^i = f_w^i / f_{w'}^i$.

The statistical hypotheses for the equivalence test are:

$$H_{0_w} : \mu_{R_w} \geq \delta \text{ or } \mu_{R_w} \leq 1/\delta \text{ vs } H_{1_w} : 1/\delta < \mu_{R_w} < \delta.$$

The ratio between two frequencies, $r_w^i = f_w^i / f_{w'}^i$, is an effect size measure. As in many studies, e.g. [13], we consider the effect to be weak when it assumes values between 1.1 and 1.3 and we explore these lower effect size values as a tolerance to conclude practical equivalence. When the sample size is high, by the central limit theorem, we use the z interval for the unknown true value of μ_{R_w}, which is,

$$(\overline{R_w} \mp z \times SE(R_w))$$

where $SE(R_w) = S_{R_w} / \sqrt{n}$.

In this case, the equivalence tests procedure consists of obtaining the confidence interval for the parameter and checking if it is contained in the interval $]1/\delta, \delta[$. If so, H_{0_w} is rejected and for the (w, w') pair, the equivalence can be assumed.

For each word length k, we construct 4^k equivalence tests. When we reject all of the 4^k null hypotheses, we consider that the symmetry phenomenon is present, as all symmetric pairs are equivalent in a global way. Since we conduct simultaneous tests, we apply the Bonferroni correction.

2.2   Correlation

We use the Pearson's correlation to measure the global symmetry effect in each individual. In particular, we use the coefficient as a score of global symmetry in each individual

$$SS_i = \frac{\sum\limits_{w}\left[(N_w^i - \overline{N^i})(N_{w'}^i - \overline{N^i})\right]}{\sum\limits_{w}\left[(N_w^i - \overline{N^i})^2\right] \sum\limits_{w}\left[(N_{w'}^i - \overline{N^i})^2\right]} \tag{1}$$

$i \in \{1, 2, ..., n\}$.

To evaluate the significance of correlation between a pair of symmetric words, we apply the one tailed Pearson correlation test. Considering ρ the Pearson correlation parameter, the tests hypothesis are:

$$H_0 : \rho = 0 \text{ vs } H_1 : \rho > 0.$$

with $T = c\sqrt{\frac{n-2}{1-c^2}} \underset{under\,H_0}{\widetilde{}} t_{n-2}$ and c the sample Pearson correlation coefficient.

2.3   Generating 1,092 Individual Human Genomes

We use the GRCh37.1 reference human genome assembly [2] and version 3 (March 16, 2012) of a Phase 1 integrated variant call set based on both low coverage and exome whole genome sequencing data from 1,092 individuals [1]. The VCF files contain the alterations necessary to incorporate in the reference human genome in order to obtain a different, individual human genome. We developed a package of custom-made C programs to generate alternate FASTA genomes from population sequencing VCF data, and to count occurrences of words from these individual genomes.

3   Results

We apply traditional statistical tests to compare the means of the occurrences of symmetric words. As expected, globally, there are significant differences for all word lengths. Also, for each word length, almost all pairs have significant differences.

As discussed in the methods section, we use equivalence testing in this analysis. Table 1 displays the percentage of equivalent pairs (in the sense of what has been described in subsection 2.1) for each k-mer length and each tolerance value (δ). We verify equivalence between symmetric pairs for $k \leq 7$, for both tolerance values $\delta = 1.1$ and $\delta = 1.3$.

Figure 1 displays an error bar plot of the global scores of symmetry (SS, equation 1). We observe high score values (close to 1) for all word lengths $k \in \{1, 2, ..., 12\}$. Note that, though all global scores of symmetry have high values, these might be attributable to the contribution of a few outliers. In this figure, we observe a high association between k and the scores (approximately parabolic behavior, concavity down with

Table 1. Percentage of equivalent tests that reject the null hypothesis

k	$\delta = 1.1$	$\delta = 1.3$
1	100	100
2	100	100
3	100	100
4	100	100
5	100	100
6	100	100
7	100	100
8	99.67	100
9	96.60	99.82
10	85.17	97.98
11	64.84	89.50
12	40.20	71.52

inflection point in $k = 4$). Hence, the global symmetry score has high values in the human genome. However, this score has a tendency to decrease as the word length increases.

For each word length k, there are 4^k pairs of symmetric words. The correlation coefficient and the corresponding statistical test p-value are obtained for each symmetric word pair based on a sample of 1092 individuals. Table 2 displays the frequency table of the correlation coefficients, highlighting the corresponding conclusion of t correlation tests. We observe that, for $k \leq 4$, the result is in accordance with the previous conclusion. However, for $k > 4$, the correlation values are very low. We also observe some not significantly positive correlations.

Table 2. Percentage of correlation coefficients in each class of effect size. *the p-value of one tailed Pearson correlation test is <0.05.

Correlation	k=1	k=2	k=3	k=4	k=5	k=6	k=7	k=8	k=9	k=10	k=11	k=12
$[-1;0[$	0	0	0	0	0.2	2.0	8.9	21.3	37.8	46.1	48.7	49.5
$[0;0.05[$	0	0	0	0	0	3.1	7.3	18.3	26.4	27.1	27.3	28.3
$[0.05;0.10[*$	0	0	0	0	0.4	2.7	6.6	18.6	18.2	15.5	14.5	13.8
$[0.10;0.30[*$	0	0	0	0	2.5	10.0	31.4	35.1	16.2	10.7	9.2	8.3
$[0.30;0.50[*$	0	0	0	0	5.5	8.8	31.1	4.8	1.0	0.3	0.2	0.1
$[0.50;1] *$	100	100	100	100	91.4	73.4	14.6	2.0	0.4	0.2	0.0	0.0

To clarify the obtained correlation values/tests for each symmetric pair, we compute, for each k, the mean, minimum, maximum and standard deviation of the correlation. Figure 2 displays these statistics for all words of length k. We observe a curious tendency that as k increases, the mean of the correlation tends to zero. Hence, the previous high values of the global symmetry score may be a consequence of only a few pairs of very frequent symmetric words.

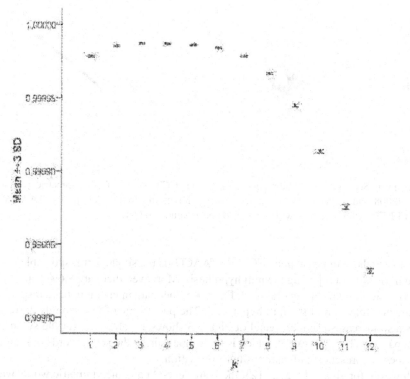

Fig. 1. Error bar of the scores of symmetry (*SS*) in 1092 human genomes

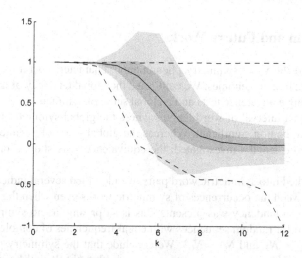

Fig. 2. Summary of statistics of the correlation coefficients between pairs of symmetry words in 1092 human genomes: mean (continuous line), minimum (dashed), maximum (dashed). The shaded region represents the standard deviation around the mean (mean ∓ standard deviation in the darker gray region and mean ∓ 3 standard deviations in the lighter gray region.

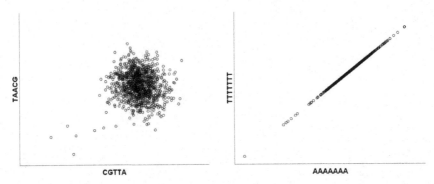

Fig. 3. Left: Scatter plot of the frequencies of the (CGTTA, TAACG) symmetric pair, with $r = -0.008$ and p value 0.790. Right: Scatter plot of the frequencies of the(AAAAAAA, TTTTTTT) symmetric pair, with $r = 0,870$ and p value < 0.001.

For $k=5$, the symmetric pair (CGTTA, TAACG) is the single pair responsible for the hypothesis test not rejecting the null hypothesis. Moreover, there are 8.6% pairs where the correlation is not strong (Table 2). For $k > 5$, there are many more pairs responsible for the non rejection of the null hypothesis. The percentage of not strongly correlated pairs also increases. The left panel of Figure 3 shows a scatter plot for the symmetric word pair (CGTTA, TAACG), which is the single pair, in the set of words of length 5, that does not present significant positive correlation.

However, for all $k \in \{1, 2, ..., 12\}$, there are several pairs of symmetric words where the correlation is significantly positive and strong. An example is displayed in the right panel of Figure 3, representing the word symmetric pair (AAAAAAA, TTTTTTT).

4   Conclusion and Future Work

Here, we studied the word symmetry phenomenon, characterized through word frequencies, in 1092 human genomes. We confirmed the global tendency of the symmetry phenomenon using equivalence tests and a global score of symmetry.

We identified an interval of word lengths where the global symmetry phenomenon tendency starts to be non significant. Whereas the global score of symmetry has high values for all word lengths investigated, the equivalence tests show a breakdown of symmetry for $k > 7$.

When we looked into symmetric word pairs, we identified several indicators of lack of similarity between the occurrences of symmetric words even when the global symmetry phenomenon tendency was present. This is surprising, as no symmetry breakdown is expected for random sequences with equal frequencies of complementary nucleotides (i.e. $N_A \sim N_T$ and $N_C \sim N_G$). We conclude that the symmetry phenomenon is less prevalent in human genomes than previously thought. It will be interesting to investigate this symmetry phenomenon for selected genomic regions.

5   Funding

This work was supported in part by *FEDER* funds through *COMPETE–* Operational Programme Factors of Competitiveness ("Programa Operacional Factores de Competitividade") and by Portuguese funds through the *Center for Research and Development in Mathematics and Applications* and the Portuguese Foundation for Science and Technology ("FCT–Fundação para a Ciência e a Tecnologia"), within project PEst-C/MAT/UI4106/2011 with COMPETE number FCOMP-01-0124-FEDER- 022690. SPG acknowledges funding from the European Social Fund and the Portuguese Ministry of Education and Science, and from project FCOMP-01-0124-FEDER-010095 of Portuguese Science Foundation.

References

1. The 1000 genomes project data release: Integrated variant call set for phase 1, version 3
2. Grch37 Reference human genome assembly
3. Albrecht-Buehler, G.: Inversions and inverted transpositions as the basis for an almost universal "format" of genome sequences. Genomics 90, 297–305 (2007)
4. Baisnée, P.-F., Hampson, S., Baldi, P.: Why are complementary DNA strands symmetric? Bioinformatics 18(8), 1021–1033 (2002)
5. Karkas, J.D., Rudner, R., Chargaff, E.: Separation of B. subtilis DNA into complementary strands. II. template functions and composition as determined by transcription with RNA polymerase. Proceedings of the National Academy of Sciences of the United States of America 60(3), 915–920 (1968)
6. Kline, R.B.: Beyond Significance testing: Reforming Data Analysis Methods in Behavioral Research. American Psychological Association (2004)
7. Kong, S.-G., Fan, W.-L., Chen, H.-D., Hsu, Z.-T., Zhou, N., Zheng, B., Lee, H.-C.: Inverse symmetry in complete genomes and whole-genome inverse duplication. PLoS One 4(11), 7553 (2009)
8. Migliorati, S., Ongaro, A.: Adjusting p-values when n is large in the presence of nuisance parameters. In: Statistics for Industry and Technology, Vienna, pp. 305–318 (September 2010)
9. Moore, D.S.: Statistics: Concepts and Controversies, 4th edn. Freeman (1997)
10. Qi, D., Jamie Cuticchia, A.: Compositional symmetries in complete genomes. Bioinformatics 17(6), 557–559 (2001)
11. Rudner, R., Karkas, J.D., Chargaff, E.: Separation of B. subtilis DNA into complementary strands, I. biological properties. Proceedings of the National Academy of Sciences of the United States of America 60(2), 630–635 (1968)
12. Rudner, R., Karkas, J.D., Chargaff, E.: Separation of B. subtilis DNA into complementary strands. III. direct analysis. Proceedings of the National Academy of Sciences of the United States of America 60(3), 921–922 (1968)
13. Thanassoulis, G., Vasan, R.S.: Genetic cardiovascular risk prediction — Will we get there? Circulation 122(22), 2323–2334 (2010)
14. Zhang, S.-H., Huang, Y.-Z.: Limited contribution of stem-loop potential to symmetry of single-stranded genomic DNA. Bioinformatics 26(4), 478–485 (2010)

A Clustering Framework Applied to DNA Microarray Data

José A. Castellanos-Garzón and Fernando Díaz

University of Valladolid, Department of Computer Science,
University School of Computer Science, Campus Maria Zambrano,
Plaza Alto de los Leones, 1 - 40005 Segovia, Spain
jantonio_cu@ieee.org, fdiaz@infor.uva.es

Abstract. This paper presents a case study to show the competence of our evolutionary framework for cluster analysis of DNA microarray data. The proposed framework joins a genetic algorithm for hierarchical clustering with a set of visual components of cluster tasks given by a tool. The cluster visualization tool allows us to display different views of clustering results as a means of cluster visual validation. The results of the genetic algorithm for clustering have shown that it can find better solutions than the other methods for the selected data set. Thus, this shows the reliability of the proposed framework.

Keywords: DNA microarray data, Data clustering, Genetic Algorithm, Data mining, Visual analytics.

1 Introduction

The study of gene expression data from DNA microarrays is of great interest for Bioinformatics (and functional genomics), because it allows us to analyze expression levels in hundreds of thousands of genes in a living organism sample. This feature makes gene expression analysis a fundamental tool of research for human health. It provides identification of new genes that are key in the genesis and development of diseases. However, the exploration of these large data sets is an important but difficult problem. The use of evolutionary and visual analytics techniques can help to cope with this problem. Visual data exploration has high potential and many applications in data mining use information visualization technology for an improved data analysis.

Classifying DNA microarray data according to their similarity degree is one of the main goals of *data mining* applied to this domain. The organization of objects in affinity groups is one way of knowledge discovery, being a key factor in *machine learning* [1,2]. Moreover, the application of *genetic algorithms* [3,4] to data mining is still of great importance in classification problems. Particularly, we can highlight, the use of evolutionary strategies to *unsupervised classification (cluster analysis)* in the knowledge discovery process [5]. Cluster analysis is one component of exploratory data analysis, which means sifting through data to make sense out of measurements by whatever means are available, whereas

M.S. Mohamad et al. (Eds.): 7th *International Conference on PACBB*, AISC 222, pp. 15–22.
DOI: 10.1007/978-3-319-00578-2_3   © Springer International Publishing Switzerland 2013

genetic algorithms (GAs) are blind search methods, inspired by the natural selection mechanism and oriented to solve complex optimization problems. Therefore, the application of evolutionary, cluster analysis and visualization techniques to DNA microarrays can turn biological data into knowledge of biological systems, often requiring further experimentation from initial data [6].

According to all previously explained, this paper presents a clustering framework joining a GA with a visualization tool, both applied to cluster analysis of DNA microarray data. The goal of this paper is to firstly show, the research in which we have been working [7,8] and secondly, the results of such a framework (particularly, for our GA-based hierarchical clustering method) for a new case study, showing the reliability of the clustering framework.

2   A Clustering Framework

This section presents details on the our framework oriented to clustering. Firstly, a GA-based hierarchical clustering has been given, after that, a tool for visual cluster analysis has also been given by showing its functionalities on DNA microarray data and finally, details on the implementation of the framework are outlined.

2.1   A Genetic Algorithm for Hierarchical Clustering

A hierarchical clustering method is a procedure for transforming a proximity matrix (matrix of distances between the objects to be grouped) into a sequence of nested partitions (clusterings), and a clustering is a type of classification imposed on a finite data set [1]. Since the hierarchical clustering is very important in biological data analysis due to its graphic representation of the data in form of *dendrogram*, which provides knowledge of the data domain, we have considered dendrograms as the individuals of our GA for hierarchical clustering.

Based then on the above, individuals (chromosomes) in our GA have been encoded as shown in Figure 1 (on the left side) for objects $\{x_1, x_2, x_3, x_4, x_5\}$. Thus, individuals are built in an agglomerative way and each level is represented by its corresponding number. That is, an individual is a dendrogram which consists of a collection of clusterings, where each clustering is a partition of the universe of objects to be clustered. At first, each dendrogram of the initial population of our method (which we call HCGA) is built up from the first level to the higher level by joining two clusters randomly chosen in the current level to create the next one [7].

Fitness Function: To measure the goodness of candidate solutions (dendrograms) of HCGA, we have based on the concepts of homogeneity and separation given in [9]. In general, we have focused on the idea of *the objects inside a cluster are very similar whereas the objects located in distinct clusters are very different*. Therefore, the fitness functions for both, clustering and dendrogram have been defined as follows:

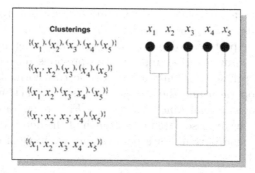

Fig. 1. Example of a hierarchical clustering on the left side and its dendrogram on the right

Definition 1. *Clustering fitness function*
Let $\mathfrak{D}$ be the proximity matrix of a given data set, then the fitness function of a clustering $\mathfrak{C}_{i+1}$ in a dendrogram $\mathfrak{G}$ (i, the level in the dendrogram) according to $\mathcal{H}_1^*$ and S_1^* (homogeneity and separation respectively) is defined as:

$$g_c(\mathfrak{C}_{i+1}) = \frac{S_1^*(\mathfrak{C}_{i+1})}{g - k + 1} - \frac{\mathcal{H}_1^*(\mathfrak{C}_{i+1})}{k - 1} + \max \mathfrak{D}, \tag{1}$$

where $S_1^*(\mathfrak{C}_{i+1})$ and $\mathcal{H}_1^*(\mathfrak{C}_{i+1})$ have been defined in [7], $k = |\mathfrak{C}_i|$ and $g = \binom{k}{2}$, being the number of distances among the clusters of $\mathfrak{C}_{i+1}$.

Definition 2. *Dendrogram fitness function*
The fitness function of a dendrogram $\mathfrak{G}$, being $\mathfrak{C}_i$ ($i \in [1, |\mathfrak{G}| - 1]$) clusterings of $\mathfrak{G}$:

$$g_d(\mathfrak{G}) = \frac{1}{|\mathfrak{G}| - 1} \sum_{i=1}^{|\mathfrak{G}|-1} g_c(\mathfrak{C}_i). \tag{2}$$

Once defined the fitness function for the individuals, our goal is to maximize g_d by obtaining small values for $\mathcal{H}_1^*$ and large values for S_1^* on the clusterings of $\mathfrak{G}$. Based then on the previous definition, an *agglomerative coefficient* (*ac*) can be used in order to estimate the level of a dendrogram $\mathfrak{G}$ where carrying out a cut-off, that is:

Definition 3. *Agglomerative coefficient*
Let $\mathfrak{G}$ and $\mathfrak{C}_i$ be a dendrogram on $\mathfrak{P}_n$ and a clustering of $\mathfrak{G}$, respectively. The agglomerative coefficient of $\mathfrak{G}$ is defined as:

$$ac(\mathfrak{G}) = \arg_{i \in [1, |\mathfrak{G}|]} \max g_c(\mathfrak{C}_i), \tag{3}$$

that is, level i whose clustering has the maximum fitness of the whole dendrogram.

Mutation Operator: We have used the mutation operator (MO) defined in [7], which is a unitary alteration operator which only is applied on a single individual. It only transforms a part of a dendrogram, exploring different branches and returning a new dendrogram that replaces the previous one. Hence, the MO carries out an in-depth search. Only a part of the transformed dendrogram is modified with this operator and the another part is kept unchangeable. Indeed, since a dendrogram is special kind of tree (see Figure 1), this MO performs similar as moving a cluster associated to a branch of the dendrogram to another branch in the same dendrogram. For a performance example of this operator as the next one, see [7].

Crossover Operator: The crossover operator (CO) has also been defined in [7] and performs by recombining valuable information of two individuals in order to build a new individual, which inheres the genetic code of their ancestors. Thus, this carries out a wide search in the dendrogram space, taking two parent dendrograms as its input to obtain a single child dendrogram. On the one hand, this CO recombines clusters of the parent dendrograms and on the other hand, it also remains other clusters unchangeable of the parents to form the child dendrogram. In general terms, the CO randomly chooses the same level in both parents to form a new clustering (which is called seed clustering) by selecting the best clusters of the parents in the chosen level. After that, the child dendrogram is built by applying the above MO on the seed clustering to achieve the upper levels of the child and finally, a divisive strategy (for clusters) is also applied on the seed clustering to build the remaining lower levels.

2.2 A Tool for Visual Cluster Analysis

Complementing our clustering framework, we have developed tool *3D-VisualCluster* (3D-VC for short) [8], which loads the results of HCGA (and in general, the results of other methods) and display them on different visualization components as a means of result visual comparison. Therefore, 3D-VC implements exploration of dendrograms (as input) in different ways: views of microarrays (heatmap) including dendrogram and parallel coordinates for clusters, 3D scatter plot (dimensionality reduction based on *principal component analysis* by using the covariance matrix) with the clusters of the data as points, which can show boundary points and 3D-shapes for clusters and additionally, a reference partition can be loaded by tool to be compared with the clusterings of the dendrogram. For this case, the clusters in the reference partition (called *r-clusters*) are displayed through their 3D-surfaces.

A general view of 3D-VC has been shown in Figure 2 displaying a sketch of eight linked and interactive views and six tasks for visual cluster analysis. Note that the order of tasks (six tasks) envolved in this figure states a methodology to follow in the visual analysis and validation of the results for a clustering method. Therefore, from the input of clustering results to 3D-VC, the user can follow the order of the tasks defined in this figure to make a process of visual analytics

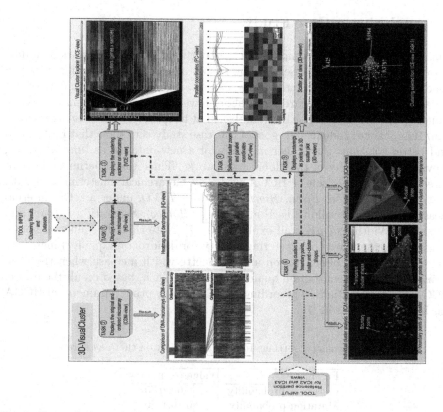

Fig. 2. General view of the tool. There are eight linked views: microarray, dendrogram and parallel coordinates views at the top of the figure; 3D scatter plot views at the bottom; cluster boundary points, reference partition surfaces and cluster surface reconstruction.

leading to knowledge discover from the interaction with the views provided by each task (see [8] for more details).

2.3   Implementing the Framework

The clustering framework has been implemented in *R-Project* (R Development Core Team) and *Java (Java-3D)*. The evolutionary part of our framework (method HCGA) has been developed on R and is freely distributed at http://cran.r-project.org/web/packages/clustergas. On the other hand, tool 3D-VC has been developed on Java and is linked to R in such as a way that the results of clustering methods implemented on R can be read by the tool for subsequent visual analysis. 3D-VC is publicly available at http://www.analiticavisual.com/jcastellanos/3DVisualCluster/3D-VisualCluster.

3 Analyzing a Case Study

This section outlines two comparative tests of the proposed framework on public data set *lung*, which comprises 73 lung tissues including 67 lung tumors for 916 gene observations for each lung tissue, namely, a gene expression matrix of 916 genes × 73 samples. 20-nearest neighbors have been used to estimate missing values of this data set. It has been published at `http://genome-www.stanford.edu/lung_cancer/adeno`. The goal of this case study is to show that HCGA can find better solutions than other methods for a new data set as *lung*, this way, proving the reliability of the global framework . Thus, the experiments are focused on the comparison of HCGA with five hierarchical clustering methods, that is, *Agnes, Diana, Eisen, HibridHclust* and *TSVQ*, under a set of internal measures of cluster validity as Homogeneity (*Homog*), Separation (*Separ*) and Silhouette Width (*SilhoW*), all explained in [7]. Note that the values of homogeneity decrease when the clustering quality (or dendrogram quality) increases, whereas the values of separation and silhouette width increase when the clustering quality (or dendrogram quality) increases. Then, based on all the above, HCGA has been initialized according to Table 1, any other parameter of HCGA has been assigned as in [7].

Table 1. Parameter settings to evaluate HCGA on the *lung* data set

Parameter	Value (or interval)
Crossover probability	$[0.60, 0.75]$
Mutation probability	$[0.10, 0.20]$
Number of individuals	20
Number of generations	$[10^3, 10^6]$
Metric on data	*Euclidean*

3.1 Global and Local Comparison of Clustering Results

The performance of HCGA with respect to other methods is shown by means of global and local quality of the found solutions. Thus, by the first case (global quality), the measures of cluster validity are applied to the whole dendrogram for each compared method. Table 2 lists the values scored by each validity measure applied to the dendrogram returned by each method, the best values reached in each case have been underlined. The second case, that is, local quality, Table 3 lists the measure values applied to the best clustering[1] of each output dendrogram. Column *#Clusters* has the number of clusters of the clustering selected for each method and the best values for each measure have also been underlined. Finally, the runtimes of HCGA defined for this experiment were between 0.30 and 2 hours.

---

[1] The best clustering according to *ac*, Definition 3.

Table 2. Global cluster validity of HCGA vs. five hierarchical clustering methods from the *lung* data set

Method	Homog	Separ	SilhoW
Agnes	14.25	20.84	<u>0.14</u>
Diana	12.79	15.31	0.05
Eisen	13.27	15.17	-0.03
HybridHclust	12.15	15.03	0.05
TSVQ	<u>12.13</u>	15.03	0.04
HCGA	14.38	<u>21.60</u>	<u>0.14</u>

Table 3. Local cluster validity of HCGA vs. five hierarchical clustering methods based on the best clustering of each output dendrogram from the *lung* data set

Method	#Clusters	Homog	Separ	SilhoW
Agnes	7	14.45	22.30	0.23
Diana	29	11.98	15.10	0.05
Eisen	9	13.40	15.19	-0.01
HybridHclust	31	<u>11.27</u>	14.90	0.03
TSVQ	31	11.31	14.90	0.03
HCGA	3	14.61	<u>23.11</u>	<u>0.30</u>

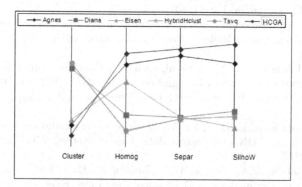

Fig. 3. A view of parallel coordinates for four measures, representing six curves with values scored by each method in Table 3 (*lung* data set)

As these two tables show, the best values for separation and silhouette width have been achieved for the HCGA and *Agnes* method. The best values for homogeneity have been achieved for methods $TSVQ$ and $HybridHclust$. Figure 3 shows the above by representing each method as a curve, where the methods in groupings $\{HCGA, Agnes\}$ and $\{HybridHclus, TSVQ\}$ have similar behavior. Summarizing on HCGA, we can say that it performs better than the other methods on separation measures and on measures that combine homogeneity and separation (such as, silhouette width), but it does not have the same performance on homogeneity. Note that as a part of our framework, the given results can visually be analyzed by tool 3D-VC to validate the used measures [8].

4   Conclusions

The goal of this paper has been to show the results of our research on a new case study. According to this, we have presented a clustering framework joining a genetic algorithm (HCGA) and a tool (3D-VC) for the analysis of hierarchical clustering. The results shown on the evolutionary part of the framework, that is HCGA, prove that it has performed well for the new used data set (*lung*), finding better solutions than the other methods. Moreover, we have found out that HCGA performs better on separation and measures that combine homogeneity and separation. On the other hand, the functionalities shown by 3D-VC can be useful in the visual validation of clustering results. Thus, our global proposal has proven its reliability for the cluster analysis of DNA microarray data.

References

1. Jain, A.K., Dubes, R.C.: Algorithms for Clustering Data. Prentice Hall, Englewood Cliffs (1998)
2. Kaufman, L., Rousseeuw, P.J.: Finding Groups in Data. An Introduction to Clustering Analysis. John Wiley & Sons, Inc., Hoboken (2005)
3. Goldberg, D.E.: Genetic Algorithms in Search, Optimization, and Machine Learning. Addison Wesley Longman, Inc. (1989)
4. Holland, J.H.: Adaptation in Natural and Artificial Systems: An Introductory Analysis with Applications to Biology, Control, and Artificial Intelligence. MIT Press Edition (1992)
5. Laszlo, M., Mukherjee, S.: A genetic algorithm that exchanges neighboring centers for k-means clustering. Pattern Recognition Letture 28, 2359–2366 (2007)
6. Bourne, P.E., Wissig, H.: Structural Bioinformatics. Wiley-Liss, Inc., Hoboken (2003)
7. Castellanos-Garzón, J.A., Díaz, F.: An evolutionary computational model applied to cluster analysis of DNA microarray data. Expert Systems with Applications 40, 2575–2591 (2013)
8. Castellanos-Garzón, J.A., García, C.A., Novais, P., Díaz, F.: A visual analytics framework for cluster analysis of DNA microarray data. Expert Systems with Applications 40, 758–774 (2013)
9. Jiang, D., Tang, C., Zhang, A.: Cluster analysis for gene expression data: A survey. IEEE Transactions on Knowledge and Data Engineering 16(11), 1370–1386 (2004)

Segmentation of DNA into Coding and Noncoding Regions Based on Inter-STOP Symbols Distances

Carlos A.C. Bastos[1], Vera Afreixo[2], Sara P. Garcia[1], and Armando J. Pinho[1]

[1] Signal Processing Lab, IEETA and Department of Electronics Telecommunications and Informatics, University of Aveiro, 3810-193 Aveiro, Portugal
{cbastos,spgarcia,ap}@ua.pt
[2] Department of Mathematics, University of Aveiro, 3810-193 Aveiro, Portugal
vera@ua.pt

Abstract. In this study we set to explore the potentialities of the inter-genomic symbols distance for finding the coding regions in DNA sequences. We use the distance between STOP symbols in the DNA sequence and a chi-square statistic to evaluate the nonhomogeneity of the three possible reading frames. The results of this exploratory study suggest that inter-STOP symbols distance has strong ability to discriminate coding regions.

Keywords: inter-STOP symbols distance, DNA, coding regions, chi-square.

1 Introduction

It is well known that DNA sequences present a nonhomogenous distribution along the sequence (e.g. coding regions have a tendency to reveal three-base periodicity [1] [2] [3]). There are many published algorithms for coding regions location (e.g. [3][4] [5] [6] [7]). However, their accuracy needs improvement [5] [6] and there is room for improvement.

In previous work, we explored the inter-nucleotide and inter-dinucleotide distance, i.e., the distance to the first occurrence of the same nucleotide (dinucleotide), to perform a comparative analysis between species [8] [9]. In this work, we extend the concept and explore the inter-STOP symbols distance over different reading frames in the DNA sequence.

It is well known that the distributions of STOP symbols in coding regions and noncoding regions are different. In the correct reading frame of coding regions the STOP symbols occur only at the end [5]. Motivated by the expectation that the distance between STOP symbols has higher values in the correct reading frame than in the other reading frames, we study, in this work, the potentiality of inter-STOP symbols distance distribution for DNA segmentation.

2 Methods

2.1 Inter-STOP Symbols Distance Sequence

The inter-STOP symbols distance sequence is a special case of the inter-trinucleotide distance sequence. It is possible to generate three trinucleotide sequences, one for each reading frame, from a single genomic sequence.

M.S. Mohamad et al. (Eds.): *7th International Conference on PACBB*, AISC 222, pp. 23–28.
DOI: 10.1007/978-3-319-00578-2_4 © Springer International Publishing Switzerland 2013

As an illustrative example consider a genomic sequence starting by

$$AAACAAACTGACACAAAACACTAATAGTTTAAAATAATAATGA\ldots.$$

Then, the three trinucleotide reading frames (R_1, R_2 and R_3) produce the following trinucleotide sequences,

R_1: *AAA CAA ACT GAC ACA AAA CAC* **TAA TAG** *TTT AAA ATA ATA ATG A* $\cdots$

R_2: *A AAC AAA CTG ACA CAA AAC ACT AAT AGT TTA AAA* **TAA TAA TGA** $\cdots$

R_3: *AA ACA AAC* **TGA** *CAC AAA ACA CTA ATA GTT* **TAA** *AAT AAT AAT GA* $\cdots$

The distance sequence for each trinucleotide is a vector containing the distances between consecutive occurrences of that trinucleotide. In this work we are interested in the inter-STOP symbols distance, i.e. the distance between consecutive stop symbols: TAA, TAG, TGA. Any of these three symbols signals the end of genes.

As an example, and using the previous nucleotide sequence, we present the beginning of inter-STOP distance sequences for each of the three reading frames:

$$d_{R_1}^{STOP} = (1, \cdots)$$
$$d_{R_2}^{STOP} = (1, 1, \cdots)$$
$$d_{R_3}^{STOP} = (7, \cdots)$$

2.2 Chi-square Statistic

We use the chi-square statistic to measure the lack of homogeneity of the inter-STOP distance distribution between the three possible reading frames. In order to compute the chi-square statistic along the trinucleotide sequences we use a sliding window of fixed length (w) in each frame, and the distances within each window are classified into 2 categories: short distance and long distance. The value used to separate the short and long distances was called cut-off (note: the long distances include the distance corresponding to the cut-off value). We also include an extra category with the number of non stop symbols within the window.

For each DNA sequence we construct contingency tables at each trinucleotide with a window of w trinucleotides. Table 1 shows the structure of the contingency tables.

Table 1. Contingency table for each window. Note: $n_{\cdot 1} = n_{\cdot 2} = n_{\cdot 3} = w$ and $n_{1j} = w - n_{2j} - n_{3j}$.

	Frame 1	Frame 2	Frame 3	Total
non STOP	n_{11}	n_{12}	n_{13}	$n_{1\cdot}$
short distance	n_{21}	n_{22}	n_{23}	$n_{2\cdot}$
long distance	n_{31}	n_{32}	n_{33}	$n_{1\cdot}$
total	$n_{\cdot 1}$	$n_{\cdot 2}$	$n_{\cdot 3}$	N

The chi-square statistic is given by

$$X^2 = \sum_i \sum_j \frac{\left(n_{ij} - \frac{n_{\cdot j} n_{i\cdot}}{N}\right)^2}{\frac{n_{\cdot j} n_{i\cdot}}{N}}.$$

When $\frac{n_{.j}n_{i.}}{N} = 0$, we consider $X^2 = 0$, with $i, j \in \{1, 2, 3\}$. This value means that the inter-STOP symbols distributions in the three reading frames are homogenous.

2.3 DNA Data

We used genomic data files obtained from the European Bioinformatics Institute site (http://www.ebi.ac.uk/genomes/) for 5 bacteria and the 16 chromosomes of *Saccharomyces cerevisiae* S288c. The bacteria were: Aster yellows witches-broom phytoplasma AYWB; *Borrelia burgdorferi* B31; *Buchnera aphidicola* (Cinara tujafilina); *Candidatus Carsonella* ruddii CE isolate Thao2000; and *Mycoplasma gallisepticum* CA06_2006.052-5-2P.

We extract the genomic sequences and the information of the position of the coding regions from the data files. This information is used to compare with the results of the chi square statistic and to evaluate its discrimination capacity.

We only considered the 5' to 3' strand and consequently we did not use the information for the genes on the complement strand.

2.4 Procedure

We obtain the chi-square statistic for each symbol of the three reading frames for a sliding window with fixed length (1000 symbols) and we vary the cut-off distance from 50 to 350 symbols. We use the ROC (receiver operating characteristic) curve and compute the area under the ROC curve (AUC) to evaluate the discrimination accuracy of the chi-square statistic. The procedures with higher AUC have better performance. Note that if the AUC is 1 the discrimination is perfect, and if the AUC is 0.5 the discrimination is worthless.

3 Results

Figure 1 shows the position of the coding regions in each of the trinucleotide reading frames and the inter-STOP symbols distances at the positions where the STOP symbols occur. As can be seen from the figure, there is a long inter-STOP distance close to the beginning of most of the contiguous coding regions in one (and only one) of the reading frames.

We used a sliding window of length 1000 (corresponding to 1000 trinucleotides) which is a reasonable compromise for the genomic sequences considered in this work. In all the genomic sequences used in this study, the percentage of contiguous coding regions whose length ≤ 1000 (trinucleotides) is at least 90%.

As mentioned previously, we varied the cut-off distance that separates the short and long distances. Table 2 shows the AUC for the various cut-off distances studied. In order to reduce the size of the table, we show only the mean and standard deviation (sd) of the AUC for the *Saccharomyces cerevisiae* chromosomes and the bacteria under study.

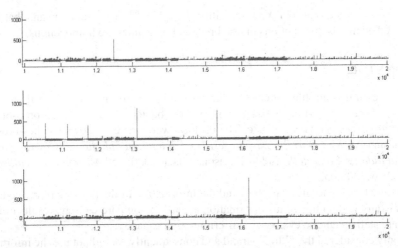

Fig. 1. Plot of the inter-STOP symbols distances for 10000 trinucleotides of the *Saccharomyces cerevisiae* chromosome I in the three frames. The coding regions are marked with thick lines.

Table 2. Area Under the ROC Curve (AUC) to discriminate coding regions using the inter-STOP distance and the chi-square statistic

Species	cut-off						
	50	100	150	200	250	300	350
	mean (sd)	mean (sd)	mean (sd)	mean (sd)	mean (sd)	mean (sd)	mean (sd)
Saccharomyces cerevisiae (16 Chr)	62% 3%	67% 2%	76% 2%	79% 1%	80% 1%	80% 1%	79% 1%
Bacteria (5)	67% 6%	79% 5%	82% 4%	82% 2%	80% 2%	78% 5%	75% 8%

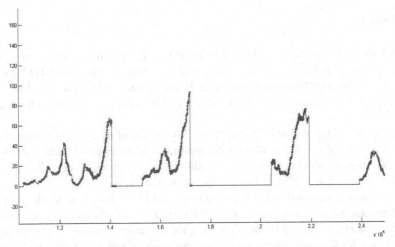

Fig. 2. Plot of chi-square values at each trinucleotide position for part of the *Saccharomyces cerevisiae* chromosome I DNA sequence. The thick lines highlight the positions corresponding to coding regions.

The discrimination capacity of the chi-square procedure varies with the cut-off distance; it has a maximum around 250 for the *Saccharomyces cerevisiae* and around 200 for the bacteria.

Figure 2 shows, as an example, the behavior of the chi-square statistic in a section of the *Saccharomyces cerevisiae* chromosome I. The coding regions are highlighted with a thick line. The method seems to have some difficulty in separating coding regions that are very close together. However, the chi-square statistic has non zero values in most of the coding regions showing heterogeneous inter-STOP distance distributions for the three reading frames .

4 Conclusion and Future Work

In this work we evaluated the possibility of the inter-STOP symbols distance for discriminating coding and noncoding regions in DNA sequences.

We conclude that the inter-STOP symbols distance combined with the chi-square statistic has potential for discriminating coding regions. We believe that this exploratory study may be extended to improve the performance of the procedure presented. In the future, we intend to study the effect of various parameters (e.g., window length, cut-off distance, number of categories in the chi-square statistic) on the discrimination accuracy of the procedure. We also intend to study the association between the coding regions lengths and the parameters of the procedure. The difficulty in separating very close coding regions may be limiting the overall quality of the results. Consequently, we intend to implement a multi-scale procedure based on chi square statistics and the variation of the window length and the cut-off distance. For eukaryotes with genes whose nucleotide number is not a multiple of 3, the algorithm will have to be improved to account for the change of phase in the 3 reading frames.

We expect that the inter-STOP symbols distance will be able to complement existing methods to increase the overall performance of gene finding algorithms.

Acknowledgments. This work was supported in part by *FEDER* funds through *COMPETE*– Operational Programme Factors of Competitiveness ("Programa Operacional Factores de Competitividade") and by Portuguese funds through the *Center for Research and Development in Mathematics and Applications* and the Portuguese Foundation for Science and Technology ("FCT–Fundação para a Ciência e a Tecnologia"), within project PEst-C/MAT/UI4106/2011 with COMPETE number FCOMP-01-0124-FEDER- 022690.

References

1. Afreixo, V.M.A., Ferreira, P.J.S.G., Santos, D.M.S.: Fourier analysis of symbolic data: A brief review. Digital Signal Processing 14(6), 523–530 (2004)
2. Frenkel, F.E., Korotkov, E.V.: Using triplet periodicity of nucleotide sequences for finding potential reading frame shifts in genes. DNA Research 16(2)
3. Abbasi, O., Rostami, A., Karimian, G.: Identification of exonic regions in dna sequences using cross-correlation and noise suppression by discrete wavelet transform. BMC Bioinformatics 12, 430 (2011)

4. Wang, W., Johnson, D.H.: Computing linear transforms of symbolic signals. IEEE Trans. Signal Processing 50(3), 628–634 (2002)
5. Nicorici, D., Astola, J.: Segmentation of DNA into coding and noncoding regions based on recursive entropic segmentation and stop-codon statistics. EURASIP Journal on Applied Signal Processing 1, 81–91 (2004)
6. Deng, S., Shi, Y., Yuan, L., Li, Y., Ding, G.: Detecting the borders between coding and noncoding dna regions in prokaryotes based on recursive segmentation and nucleotide doublets statistics. BMC Genomics 13(suppl. 8), S19 (2011)
7. Tsonis, A.A., Kumar, P., Elsner, J.B., Tsonis, P.A.: Wavelet analysis of DNA sequences. Phys. Rev. E 53(2), 1828–1834 (1996)
8. Afreixo, V., Bastos, C.A.C., Pinho, A.J., Garcia, S.P., Ferreira, P.J.S.G.: Genome analysis with inter-nucleotide distances. Bioinformatics 25(23), 3064–3070 (2009)
9. Bastos, C.A.C., Afreixo, V., Pinho, A.J., Garcia, S.P., Rodrigues, J.M.O.S., Ferreira, P.J.S.G.: Inter-dinucleotide distances in the human genome: an analysis of the whole-genome and protein-coding distributions. Journal of Integrative Bioinformatics 8(3), 172 (2011)

Assignment of Novel Functions to *Helicobacter pylori* 26695's Genome

Tiago Resende, Daniela M. Correia, and Isabel Rocha

IBB – Institute for Biotechnology and Bioengineering, Centre of Biological Engineering,
University of Minho, Portugal
tiagoresende@ceb.uminho.pt,
{danielacorreia,irocha}@deb.uminho.pt

Abstract. *Helicobacter pylori* is a pathogenic bacterium that colonizes the human epithelia, causing duodenal and gastric ulcers as well as gastric cancer. The genome of *H. pylori* 26695 has been sequenced and annotated. In addition, two genome-scale metabolic models have been developed. In order to maintain accurate and relevant information on coding sequences (CDS) and to retrieve new information, the assignment of new functions to *Helicobacter pylori* 26695's genes was performed. The use of software tools, on-line databases and an annotation pipeline for inspecting each gene allowed the attribution of validated E.C. numbers to metabolic genes, and the assignment of 177 new functions to the CDS of this bacterium. This information provides relevant biological information for the scientific community dealing with this organism and can be used as the basis for a new metabolic model reconstruction.

Keywords: *Helicobacter pylori*, Genome annotation, Metabolic functions, Genome-Scale Reconstruction.

1 Introduction

Helicobacter pylori, first cultivated in 1982 [1], is a gram-negative, spiral-shaped bacterium that belongs to the Proteobacteria [2, 3]. It is well known that this bacterium colonizes the stomach of more than 50% of the human population worldwide, reaching 80% of infection rate in developing countries [1, 3]. When in the gastric mucosa, this bacterium induces a chronic inflammation causing an increase in the risk of developing a disease such as duodenal and gastric ulcer, gastric cancer and mucosa associated lymphoid tissue (MALT) lymphoma [1]. However, only few individuals develop any *H. pylori* related gastric disease [3]. This may be due to fact that this bacterium possesses mechanisms to increase genomic diversity yielding multiple and diverse strains [4]. At the present time, there are 43 completely sequenced genomes of different *H. pylori* strains on NCBI, which highlights this bacterium genetic variability. *H. pylori* 26695, a highly pathogenic strain, was originally isolated from a patient in the United Kingdom with gastritis and had its complete genome sequenced and published in 1997 using whole-genome random sequencing [5]. This organism presents a small size genome of around 1.67 Mbp, with approximately 1590 coding sequences (CDS) identified [5].

M.S. Mohamad et al. (Eds.): *7th International Conference on PACBB*, AISC 222, pp. 29–36.
DOI: 10.1007/978-3-319-00578-2_5   © Springer International Publishing Switzerland 2013

The genome functional annotation can be seen as the process of allocating functional information to the genes of a sequenced genome. The majority of gene functions are assigned by homology search from characterized sequences, found in several online databases; and, if a given gene product is unknown, it is labeled as hypothetical protein [6]. The re-annotation can be viewed as the process of updating the functional information of a genome. Databases and computational methods are constantly evolving and over time new information is also being published, making possible to assign new gene functions [7]. The last re-annotation of *H. pylori* 26695 was published in 2003. This re-annotation generated a specific database for *H. pylori* (PyloriGene) [8] and allowed the reduction of the percentage of hypothetical proteins from approximately 40% to 33%, allowing also the reassignment of functions to 108 CDS [8]. Unfortunately, this re-annotation does not contemplate the allocation of E.C. numbers to the annotated metabolic genes and therefore it compromises some of the applications of the annotation. A very important application of gene functional annotation is the reconstruction of the metabolic network of a sequenced organism. This reconstruction allows the development of a genome-scale metabolic model based on the well-known stoichiometry of biochemical reactions catalyzed by the enzymes encoded in the annotated genes of an organism [9, 10]. These models can then be used for simulating *in silico* the phenotypic behavior of a microorganism under different environmental and genetic conditions, thus representing an important tool in metabolic engineering design and the identification of novel drug targets for pathogens [10].

To date, two metabolic models of *H. pylori* 26695 were published. The model *i*CS291 was published in 2002 and contains 291 genes and 388 reactions [11]; in 2005, based on the previous model, a new model was reconstructed, the *i*IT341 GSM/GPR with 341 genes and 476 reactions, including also 355 gene-protein reaction associations [12]. Most of improvements made in the latter model were a result of the increase of available literature and the revised annotation of the *H. pylori* genome [12].

Here we present a new re-annotation of the *H. pylori* 26695 genome. The function of each gene previously annotated was reevaluated, new functions were identified and EC numbers were assigned to genes with metabolic functions, thus presenting the combined results of updated databases and new annotation methodologies. This re-annotation will be used as the basis for reconstructing an updated genome-scale metabolic model for *H. pylori* 26695.

2   Methods

H. pylori 26695's genome was retrieved, in the amino acid fasta format, from the GenBank repository at ftp://ftp.ncbi.nih.gov/genomes/Bacteria/Helicobacter_pylori_26695_uid57787/.

Merlin
merlin (MEtabolic model Reconstruction using genome scaLe INformation) is a software tool created in our group to assist on the processes of (re) annotation and

reconstruction of genome-scale metabolic models. *merlin* is available for download at http://www.merlin-sysbio.org. It performs automatic genome-wide functional (re)annotations and provides a numeric confidence score for each automatic assignment, taking into account the frequency and taxonomy within the annotation of all similar sequences [13]. In the present work the confidence score was kept with the default parameters, with a set threshold of 0.7. To perform homology searches, *merlin* uses both BLAST (Basic Local Alignment Search Tool) (from NCBI) and profile HMM (Hidden Markov Models) (from HMMER [14]) algorithms. *merlin*'s interface was used throughout the re-annotation process to assign functions to each, gene based on the highest confidence scores [13].

Annotation Pipeline

After the automatic re-annotation performed by *merlin*, each candidate function was manually inspected by following several confirmation steps as described in Fig.1. For this, three on-line databases were used: UniProt [15] which contains up-to-date information in many *H. pylori* protein coding genes; BRENDA [16] which is an enzyme curated information database, used to confirm gene product names of a certain E.C. number; and PyloriGene [8] the specific *H. pylori* annotation database released in January 2003 upon the last re-annotation of the strain 26695 and last updated, through blastp homology search, in March, 2011[8].

The manual curation of *merlin* results began with the correspondence between each candidate and the information on different databases, giving priority to Uniprot reviewed information, followed by Uniprot unreviewed and finally the information in PyloriGene. When a match with reviewed information occurred, *merlin* candidates were annotated with a very high confidence level. On the other hand, when there was a match with Uniprot unreviewed data, the candidates were annotated with high or medium confidence levels, according to the type of information present, such as E.C. numbers, for example. If there was no information on Uniprot for candidates, *merlin* homology data, PyloriGene annotation and relevant bibliographic references (if existent on PyloriGene) were analyzed. Results with the best scores were selected and annotated with high, medium or low confidence levels, according to bibliographic evidence. When mismatches occurred between *merlin* results and Uniprot, *merlin* homology results were analyzed to search for matching information, or this was manually added. Each of the potential enzyme encoding candidates was revised in BRENDA to verify its function and confirm E.C. number assignment. Some of the enzymes were assigned with incomplete EC numbers; thus, this database was also used to identify complete EC numbers when available, by searching for enzyme's product name. As previous annotation lacked E.C. number information, and due to the importance of this kind of information, for instance in the reconstruction of metabolic models, an effort was made to try to retrieve every possible E.C. number belonging to candidates encoding enzymes. Therefore, each of the potential enzyme encoding candidates' E.C. number was sought in the different sources of information, including Uniprot, *merlin* results and BRENDA. Nevertheless, despite following the annotation pipeline, genes with no metabolic function, naturally, were not assigned with an E.C. number and therefore were annotated according to the source of information, whether it was Uniprot reviewed, unreviewed, PyloriGene or *merlin* homology data.

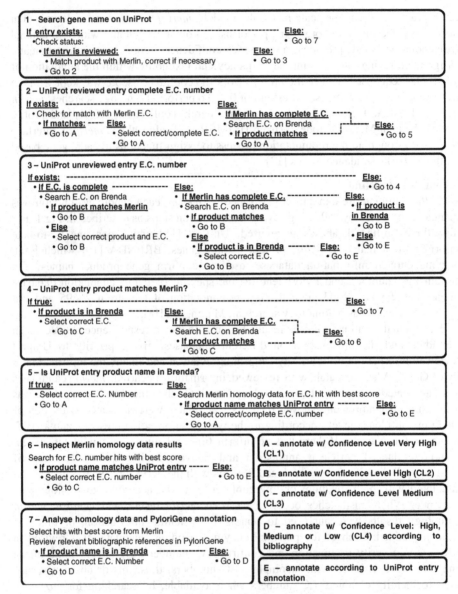

Fig. 1. Re-annotation pipeline for manual inspection of each gene candidate

3   Results and Discussion

All protein encoding genes present in *H. pylori* 26695 genome were annotated according to the proposed methodology and reviewed by the developed annotation pipeline. The number of genes inspected was different from the last re-annotation because, in the genome retrieved from NCBI, the number of genes has been updated, having now 1573 genes, instead of the previous 1590.

Function Assignment

Analyzing the results obtained from homology search with *merlin*, it was noticed that new assigned functions were based in homology with other *H. pylori* strains. This might be due to the exponential amount of *H. pylori* strains being sequenced in recent years, increasing the volume of available information on their genome.

As depicted in Table 1, the total number of coding sequences (CDS) annotated with a function was 1203, divided into 581 metabolic CDS (528 with complete E.C. numbers and 53 with incomplete E.C. numbers) and 622 non-metabolic CDS. The number of hypothetical CDS was 370, representing a total of 24% of the CDS in the genome, a lower number than the previous annotation which contained 510 hypothetical proteins (32%).

Comparing the present annotation with the previous one, it is possible to observe that both annotations are in agreement in the function of 1026 CDS. The number of CDS with differently assigned functions is 177, of which 137 correspond to the allocation of new functions to previous hypothetical CDS and 40 to the assignment of new functions to CDS previously annotated with another function. The 177 new functions assigned are divided in 71 metabolic CDS, 77 non-metabolic CDS and 29 CDS with a generic function, which has a lower level of specificity, such as, for example: HP1234, a membrane transport protein. In more than half of cases the new assignment of a function is related to increasing specificity of the function previously assigned and not necessarily to a modification in the function. For instance, the protein encoding gene HP1450, which had been annotated as an "inner membrane protein", is now assigned as a "Membrane integrase YidC".

Table 1. Distribution of CDS according to functional category

		This work	*PyloriGene*
Total CDS		1573	1590
Metabolic CDS	Complete E.C.	528	
	Incomplete E.C.	53	1080
Non-metabolic CDS		622	
Hypothetical CDS		370	510

Annotation Confidence Level

As a result of inspecting CDS according to the annotation pipeline, an annotation confidence level has been attributed to each protein coding sequence, according to the robustness of the information generated. Table 2 presents the confidence level for the total CDS and CDS with new functions.

Table 2. Confidence levels of function assignments to total CDS and CDS with new functions

Confidence Level	Total CDS (1573)	New functions (177)
Very high	529	39
High	93	4
Medium	65	6
Low	886	128

For a total of 1573 CDS annotated, 529 (33.6%) were classified with a very high confidence level, which is the highest classification, indicating that these genes are reviewed on Uniprot, and, therefore well characterized and curated manually by experts. This is also true for the 39 new functions (22%) classified in the same way. "High", is the classification level of 93 (6%) of total CDS and 4 (2%) of new functions, which, along with the medium confidence level (65 (4%) of total CDS and 6 (3%) of new functions) also indicates a good/average confidence in the results, although in a lesser extent. This classification was assigned to genes with high similarity with other genes well characterized. The majority of the total CDS, 886 (56%), and of new functions, 128 (72%) were assigned with a low confidence level, indicating that these genes are not well characterized, lacking reviewed information and validation, being the result of pure homology search data and inference methodologies. This outcome was, somewhat, expected for new functions, once new homology information is more rapidly generated than direct biological/biochemical experimental data and also because the revision, by experts, of all existing information is a laborious and time consuming task.

Enzyme Class Distribution

More than 88% (513) of the CDS assigned with metabolic activities were classified with only one complete E.C. number (monofunctional). Nevertheless, two other groups appeared, depending on the number and class of assigned E.C. number. As depicted in table 3, most of complete monofunctional E.C. numbers are classified as transferases, 163 (28%) CDS. On the other hand, most of the CDS encoding incomplete E.C. numbers are hydrolases, 23 (4%). Nevertheless, only 9% (53) of enzymes have an incomplete E.C. number. Oxireductases, transferases and hydrolases represent more than 75% of the identified enzymes. Multifunctional genes encode for more than one enzymatic function within the same class, but with different functions. They catalyze similar reactions using substrates with small differences.

Table 3. Enzyme encoding genes classification

	Complete E.C.			Incomplete E.C.		
	A[1]	B[2]	C[3]	A[1]	B[2]	C[3]
Oxidoreductases	94	1	1	2	0	0
Transferases	163	5	2	19	0	0
Hydrolases	128	2	1	23	0	0
Lyases	45	1	1	1	0	1
Isomerases	32	0	1	4	0	0
Ligases	51	0	0	3	0	0

1- A = Monofunctional; 2- B = Multifunctional; 3- C = Multiclass.

For instance, HP0683, a bifunctional N-acetylglucosamine-1-phosphate uridyltransferase/glucosamine-1-phosphate acetyltransferase (2.3.1.157, 2.7.7.23) catalyzes a reaction where the product of the first is the substrate of the second. Multiclass genes encode for more than one enzymatic activity whose E.C. numbers belong to different classes, meaning they have dissimilar catalytic functions, as for example, HP0326 which encodes for pseudaminic acid cytidylyltransferase and UDP-2,4-diacetamido-2,4,6-trideoxy-beta-L-altropyranose hydrolase (5.3.1.24, 4.1.1.48) that are classified as a transferase and a hydrolase, respectively. For constructing table 3, when classifying a protein coding sequence with more than one E.C. number, such CDS was assigned to the subgroup of first enzyme annotated, because such function was assumed as the main function.

4   Conclusions

In the present work, the assignment of new functional activities to the CDS of *H. pylori* 26695 genome was performed. Using a software tool for re-annotation and an annotation pipeline, all gene functions were inspected and updated, when necessary, being assigned with a confidence level for their function. The E.C. numbers for all metabolic CDS were searched, validated and attributed when found.

A total of 177 new functions were assigned, 137 of which were attributed to CDS previously classified as "hypothetical proteins". 40 new functions were assigned to CDS already annotated; many of them had been classified with only generic annotations. From the new functions assigned, 71 were metabolic, 77 non-metabolic and 29 had generic descriptions, indicating, for instance the localization of the protein. A total of 581 E.C. numbers were assigned to CDS, being 528 of them complete E.C. numbers. These results bring new and more comprehensive information to the *H. pylori* 26695 genome, increasing and improving the existing knowledge on this human pathogen, with special relevance for the attributed metabolic functions. The assignment of E.C. numbers is a fundamental task, since these data can be used for the reconstruction of a new genome-scale metabolic model for this organism.

Acknowledgements. This work was supported by the project FCOMP-01-0124-FEDER-009707, entitled "HeliSysBio – molecular Systems Biology in *Helicobacter pylori*" (Ref.: FCT PTDC/EBB-EBI/104235/2008). Daniela Correia is grateful for financial support from the FCT (PhD grant: SFRH/BD/47596/2008).

References

1. Marshall, B.: Helicobacter pylori: 20 years on. Clinical Medicine 2, 147–152 (2002)
2. Ge, Z., Taylor, D.E.: Contributions of genome sequencing to understanding the biology of *Helicobacter pylori*. Annu. Rev. Microbiol. 53, 353–387 (1999)
3. Kusters, J.G., Van Vliet, A.H.M., Kuipers, E.J.: Pathogenesis of *Helicobacter pylori* infection. Clinical Microbiology Reviews 19, 449–490 (2006)

4. Costa, A.C., Figueiredo, C., Touati, E.: Pathogenesis of *Helicobacter pylori* Infection. Helicobacter 14, 15–20 (2009)
5. Tomb, J.-F., White, O., Kerlavage, A.R., Clayton, R.A., et al.: The complete genome sequence of the gastric pathogen *Helicobacter pylori*. Nature 388, 539–547 (1997)
6. Dias, O., Gombert, A.K., Ferreira, E.C., Rocha, I.: Genome-wide metabolic (re-) annotation of Kluyveromyces lactis. BMC Genomics 13, 517 (2012)
7. Médigue, C., Moszer, I.: Annotation, comparison and databases for hundreds of bacterial genomes. Research in Microbiology 158, 724–736 (2007)
8. Boneca, I.G., De Reuse, H., Epinat, J.C., Pupin, M., Moszer, I., De Reuse, H., Labigne, A.: A revised annotation and comparative analysis of *Helicobacter pylori* genomes. Nucleic Acids Research 31, 1704–1714 (2003)
9. Durot, M., Bourguignon, P.-Y., Schachter, V.: Genome-scale models of bacterial metabolism: reconstruction and applications. FEMS Microbiology Reviews 33, 164–190 (2009)
10. Rocha, I., Förster, J., Nielsen, J.: Design and Application of Genome-Scale Reconstructed Metabolic Models. In: Gerdes, S.Y., Ostermnan, A.L. (eds.) Methods in Molecular Biology. Gene Essentiality, vol. 416, pp. 409–433. Humana Press Inc. (2007)
11. Schilling, C.H., Covert, M.W., Famili, I., Church, G.M., Edwards, J.S., Palsson, B.Ø.O.: Genome-scale metabolic model of *Helicobacter pylori* 26695. Journal of Bacteriology 184, 4582–4593 (2002)
12. Thiele, I., Vo, T.D., Price, N.D., Palsson, B.Ø.: Expanded Metabolic Reconstruction of *Helicobacter pylori* (iIT341 GSM/GPR): an In Silico Genome-Scale Characterization of Single- and Double-Deletion Mutants. Society 187, 5818–5830 (2005)
13. Dias, O., Rocha, M., Ferreira, E.C., Rocha, I.: Merlin: Metabolic Models Reconstruction using Genome-Scale Information. In: Banga, J.R., Bagaerts, P., Impe, J., Van Dochain, D., Smets, I. (eds.) Proceedings of the 11th International Symposium on Computer Applications in Biotechnology (CAB 2010), Leuven, Belgium, pp. 120–125 (2010)
14. Finn, R.D., Clements, J., Eddy, S.R.: HMMER web server: interactive sequence similarity searching. Nucleic Acids Research. Web Server Issue 37, W29–W37 (2011)
15. The UniProt Consortium: Reorganizing the protein space at the Universal Protein Resource (UniProt). Nucleic Acids Research 40, D71–D75 (2012)
16. Scheer, M., Grote, A., Chang, A., Schomburg, I., Munaretto, C., Rother, M., Söhngen, C., Stelzer, M., Thiele, J., Schomburg, D.: BRENDA, the enzyme information system in 2011. Nucleic Acids Research 3, 670–676 (2011)

Analysing Quality Measures of Phasing Algorithms in Genome-Wide Haplotyping

Sergio Torres-Sánchez, Manuel García-Sánchez, Nuria Medina-Medina,
and María Mar Abad-Grau

Department of Computer Languages and Systems - CITIC
University of Granada, Granada, Spain
{sergiot,mgarcia,nmedina,mabad}@ugr.es
http://lsi.ugr.es/

Abstract. Inferring haplotype phase is a subject of interest, and given the increasing number of GWAS and related applicactions, like genetic profiling, it is necessary for phasing algorithms to provide a good performance even when processing large DNA fragments. Most of studies focus on genomic regions of limited length, therefore, we propose to test the most common statistics with genetic regions of variable length. We have found that one of the most used, the switch error, is insufficient when considering long distances: it converges to a constant value which does not truly shows the quality of the inferred phase. Furthermore, the IGP (incorrect genotype percentage) is a much more precise measure of the quality of the algorithm. New phasing algorithms should not care only about the number of switches, because in some cases (classifiers to assess genetic risks, for example) is important to distinguish the haplotype of each parent to obtain better results.

Keywords: haplotype, phasing algorithms, switch errors, genetic maps.

1 Introduction

Genome-Wide Association Studies (GWAS) have experimented a great success in the recent years, specially as a tool to assess the genetic factors that influence human health. However, GWAS data usually comes under the form of unphased genotypes, while it would be interesting to reconstruct the corresponding pair of haplotypes from them, since haplotype-based models have proven to have higher accuracy [7].

Haplotype phase can be obtained by molecular methods, but it is a fairly expensive process. Instead, haplotypes are usually obtained through statistical methods, which are called computational phasing.

There are many examples of phasing algorithms in the literature, but most of them focus on reconstructing haplotypes in short ranges, assuming that most applications will work well if the haplotypes are correct locally, and do not consider to maintain the consistency between haplotypes over long distances [2].

The purpose of this paper is to show the effects of the deterioration of haplotypes over thousands of loci, and to question the measures currently used to test

M.S. Mohamad et al. (Eds.): *7th International Conference on PACBB*, AISC 222, pp. 37–44.
DOI: 10.1007/978-3-319-00578-2_6   © Springer International Publishing Switzerland 2013

the quality of phasing algorithms when phashing needs to be performed between long-distance loci.

2   Methodology

2.1   Data Processing

To prove our hypothesis, we conducted a series of tests on a group of trio samples (nuclear families) with their haplotypes resolved. In the first place, we randomly imputed the phase at each heterozygous position, i.e. for each trio offspring x_i and each heterozygous locus $l_j = Aa$, the major allele A was classified as belonging to the left individual haplotype with probability 0.5, to obtain a set of artificially unphased genotypes. Next, we obtained the phase of these individuals with a known genetic analysis software package, BEAGLE [3], first considering them as unrelated individuals, and then with trio information.

Among the different criteria to check the quality of the phasing proccess, we selected two: switch error rate (SER) and incorrect genotype percentage (IGP) [4][5]. SER is defined as the quotient between the number of switches needed to recover the original haploid sequence and the total number of possible switches, and IGP is the quotient between the number of genotypes that have their phase incorrectly inferred and the number of heterozygous loci (to calculate this percentage, it is necessary to align the estimated haplotypes with the true ones, to minimize the differences). Given $N_switchs$ the number of actual switches, N_{het} the number of heterozygous loci and N_{incGen} the number of incorrect genotypes, this measures are defined as:

$$SER = \frac{N_{switchs}}{N_{het}-1} \qquad\qquad IGP = \frac{N_{incGen}}{N_{het}}$$

We disregarded other statistics, since they are related to missing data or linkage disequilibrium, subjects not considered in this work. Thus, for each of the data sets (unphased, phased as unrelated and phased as trios), we calculated SER and IGP for fragments of genotypes or variable length, using windows of size $w = 50, 100, 500, 1000, 5000, 10000$ and a final set covering all available loci for each individual. The arithmetic mean obtained in each case is reflected in the Table 1.

To visually check the behavior of these statistics, we also built two heat maps (Figure 1). In both maps, the axes represent positions along the genome. In the map (a), the pixel x, y represents the number of genotype errors between x and y positions, divided by the total number of heterozygous positions that there are between them, i.e. the IGP. In a similar way, the map (b) shows the SER between two positions.

2.2   Datasets

We have run the tests with phased data obtained from the Phase III of the International HapMap Project [8]. More precisely, the table 1 have been generated

Table 1. Switch errors and IGP on the first chromosome for the individuals phased with BEAGLE

	Population	Unphased		Phased as unrelated		Phased as trios	
		Switch errors	IGP	Switch errors	IGP	Switch errors	IGP
$w = 50$	ASW	0.47	0.37	0.27	0.27	0.01	0.01
	CEU	0.42	0.32	0.04	0.06	0.00	0.00
	MEX	0.42	0.33	0.06	0.09	0.00	0.00
	MKK	0.47	0.36	0.10	0.14	0.00	0.00
	YRI	0.47	0.36	0.06	0.09	0.00	0.00
$w = 100$	ASW	0.49	0.41	0.27	0.33	0.01	0.01
	CEU	0.47	0.38	0.04	0.10	0.01	0.01
	MEX	0.47	0.39	0.07	0.14	0.00	0.00
	MKK	0.49	0.41	0.10	0.20	0.00	0.00
	YRI	0.49	0.41	0.06	0.15	0.00	0.00
$w = 500$	ASW	0.50	0.47	0.27	0.42	0.01	0.01
	CEU	0.50	0.46	0.04	0.26	0.00	0.00
	MEX	0.50	0.46	0.07	0.30	0.00	0.00
	MKK	0.50	0.47	0.09	0.34	0.00	0.00
	YRI	0.50	0.46	0.06	0.30	0.00	0.00
$w = 1000$	ASW	0.50	0.48	0.27	0.44	0.01	0.01
	CEU	0.50	0.47	0.04	0.33	0.00	0.00
	MEX	0.50	0.47	0.07	0.36	0.00	0.00
	MKK	0.50	0.48	0.09	0.38	0.00	0.00
	YRI	0.50	0.48	0.06	0.36	0.00	0.00
$w = 5000$	ASW	0.50	0.49	0.27	0.47	0.01	0.01
	CEU	0.50	0.49	0.04	0.42	0.00	0.00
	MEX	0.50	0.49	0.07	0.44	0.00	0.00
	MKK	0.50	0.49	0.09	0.45	0.00	0.00
	YRI	0.50	0.49	0.06	0.43	0.00	0.00
$w = 10000$	ASW	0.50	0.49	0.27	0.48	0.01	0.01
	CEU	0.50	0.49	0.04	0.44	0.00	0.00
	MEX	0.50	0.49	0.07	0.46	0.00	0.00
	MKK	0.50	0.49	0.09	0.46	0.00	0.00
	YRI	0.50	0.49	0.06	0.45	0.00	0.00
$w = 116415$	ASW	0.50	0.50	0.27	0.49	0.01	0.01
	CEU	0.50	0.50	0.04	0.48	0.00	0.00
	MEX	0.50	0.50	0.07	0.49	0.00	0.00
	MKK	0.50	0.50	0.09	0.49	0.00	0.00
	YRI	0.50	0.50	0.06	0.48	0.00	0.00

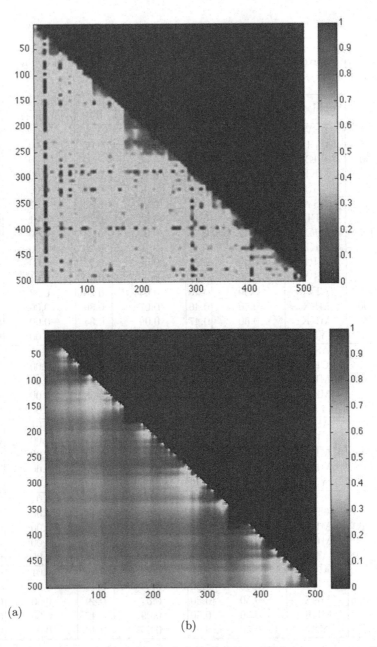

Fig. 1. Heat maps representing the statisticals between 500 heterozygous positions of a CEU individual. While (a) shows the IGP between two positions, the map (b) shows the switch error between them.

using 10 trios from southwest USA with African ancestry (ASW), 44 European trios (CEU), 23 trios from Los Angeles with Mexican ancestry (MEX), 28 Maasai trios from Kinyawa, Kenya (MKK) and 50 Yoruba trios from Ibadan, Nigeria (YRI). For each of this individuals, we have used only the first chromosome sample, composed by 116415 SNPs. Regarding the heat maps, they have been built by using 500 heterozygous positions of a random CEU individual.

3 Results

Table 1 shows that each dataset has a very different behavior. Unphased data presents a constant SER and IGP: they are always close to 50%, and and they barely vary with the window size. This makes perfectly sense because we are talking about data that have been been randomly shuffled. On the other hand, the data phased with family information presents values that are always zero or very close to zero, this makes sense too since it is a known fact that it is easy to resolve the phase almost perfectly when having family trios. However, trios are not always available, so it is necessary to improve the phasing of unrelated individuals. Therefore, we have represented the statistics of the individuals phased as unrelated in the graph of Figure 2, since they are the most interesting dataset for our work.

It is important to notice that while the SER remains constant, regardless of the window size, the IGP will gradually converge to 50%. The reasons that make

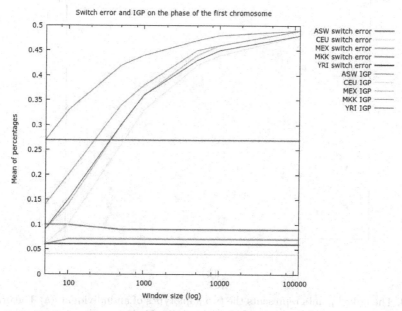

Fig. 2. Switch error rate and IGP for the populations, after phasing them with BEAGLE as unrelated individuals. The window size is represented on the x axis, on a logaritmic scale.

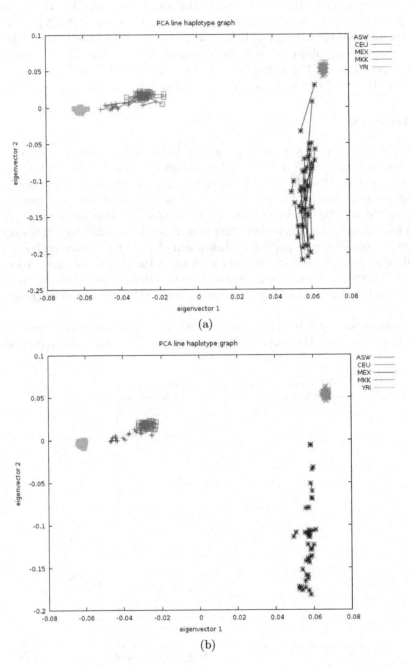

Fig. 3. The linked points represents the two haplotypes of an individual. (a) The original HapMap phased data, using trio information. (b) The former data, after unphasing it randomly and then phasing it again with BEAGLE, as unrelated individuals.

us believe that this is not the best statistic to assess the quality of a phasing algorithm are (1) the SER does not vary as the number of considered positions grows, so it does not provide too much information and (2) even if the SER has very low values, the fact that the IGP comes close to 50% means that, although the phase could be resolved locally, actually the obtained haplotypes are a mixture in the same proportion of the original haplotypes when long-distant haplotypes are considered. This behavior is reflected in Figure 1.

The map (a) has darker (and therefore, lower) IGP values on the diagonal, where the positions are close, but as we go far from this diagonal, almost every value in the graph is next to 0.50, same as in Table 1. In the map (b), there are spots colored in dark blue and bright red, representing the regions where the phase has been resolved locally and the places in which there is a switch (notice that while the blue spots are bigger, the red ones are just little dots). But the rest of the map (b) has an uniform color: this is the reason why the calculated average switch error remains constant.

We can verify that the phase is not resolved correctly by using the phased data in some analysis, and checking the quality of the results they generate. For example, in a parallel work, we used a variant of the genetic maps employed by other authors [6] in their papers: We applied a principal component analysis to the haplotypes of the parents of family trios, reducing them to two coordinates, and finally we represented the children as the segment linking the represented haplotypes of both parents (subsequently these graphics are overlayed on geographical maps, in an attemp to find matches; this work is currently submitted and pending for revision). In the Figure 3, the graph (a) corresponds to the representation of the HapMap populations with their phase resolved correctly, while graph (b) contains the same individuals, with their haplotypes shuffled randomly and then phased with BEAGLE, considering unrelated individuals. The shortcomings are quite obvious: even if the switch error rate is low, the haplotypes obtained are a mixture at a 50% rate of the original haplotypes.

We can find another example of this loss of accuracy in recent publications: the classifiers which assess the genetic risk that an individual present to a complex disease can use haplotypes too, and it has been proved that it is necessary to take into account which are the maternal and the paterna haplotype, because a random permutation will drop the quality of the results [1].

4   Conclusions

These results confirm us that the phase algorithms that does not use family information from trios, suffer a significant loss of quality, and that the switch error rate, despite being widely used, is not the measure that best reflects these shortcomings, especially when considering whole GWAS studies.

In order to obtain more accurate outputs in the analysis that make use of haplotype information, the phase must be correctly resolved, also over long distances, and not just locally. In that aspect, the IGP seems a more informative statistic about the phase resolution. Thus, new phasing algorithms should consider these results, since nowadays there are a lot of applications that need to work with large regions of the genome, such as the classifiers to assess the genetic risk to complex diseases.

Acknowledgments. The authors were supported by the Spanish Research Program under project TIN2010-20900-C04-1, the Andalusian Research Program under project P08-TIC-03717 and the European Regional Development Fund (ERDF).

References

1. Abad-Grau, M.M., Medina-Medina, N., Masegosa, A.R., Moral, S.: Haplotype-based Classifiers to Predict Individual Susceptibility to Complex Diseases - An Example for Multiple Sclerosis. Biostec Bioinformatics, 360–366 (2012)
2. Browning, S.R., Browning, B.L.: Rapid and Accurate Haplotype Phasing and Missing-Data Inference for Whole-Genome Association Studies By Use of Localized Haplotype Clustering. American Journal of Human Genetics 81(5), 1084–1097 (2007), doi:10.1086/521987
3. Browning, B.L., Browning, S.R.: A unified approach to genotype imputation and haplotype phase inference for large data sets of trios and unrelated individuals. Am. J. Hum. Genet. 84, 210–223 (2009)
4. Higasa, K., Kukita, Y., Kato, K., Wake, N., Tahira, T., et al.: Evaluation of Haplotype Inference Using Definitive Haplotype Data Obtained from Complete Hydatidiform Moles, and Its Significance for the Analyses of Positively Selected Regions. PLoS Genet. 5(5), 1000468 (2009), doi:10.1371/journal.pgen.1000468
5. Marchini, J., Cutler, D., Patterson, N., Stephens, M., Eskin, E., Halperin, E., Lin, S., Qin, Z.S., Munro, H.M., Abecasis, G.R., Donnelly, P.: A Comparison of Phasing Algorithms for Trios and Unrelated Individuals. American Journal of Human Genetics 78(3), 437–450 (2006), doi:10.1086/500808
6. Novembre, Johnson, T., Bryc, K., Kutalik, Z., Boyko, A.R., Auton, A., Indap, A., King, K.S., Bergmann, S., Nelson, M.R., Stephens, M., Bustamante, C.D.: Genes mirror geography within europe. Nature 456(7218), 98–101 (2008), doi:10.1038/nature07331
7. Torres-Sánchez, S., Medina-Medina, N., Montes-Soldado, R., Masegosa, A.R., Abad-Grau, M.M.: Riskoweb: Web-based genetic profiling to complex disease using genome-wide snp markers. In: Rocha, M.P., Corchado, J.M., Fdez-Riverola, F., Valencia, A. (eds.) Proceedings of the 5th International Conference on Practical Applications of Computational Biology & Bioinformatics (PACBB 2011), vol. 1, p. 18 (2011)
8. T. I. HapMap-Consortium, Integrating common and rare genetic variation in diverse human populations. Nature 467(7311), 52–58 (2010), doi:10.1038/nature09298

Search Functional Annotations Genetic Relationships of Coffee through Bio2RDF

Luis Bertel-Paternina[3], Luis F. Castillo[1], Alvaro Gaitán-Bustamente[2],
Narmer Galeano-Vanegas[2], and Gustavo Isaza[1]

[1] Universidad de Caldas. Manizales, Colombia
Departamento de Sistema e Informática
{luis.castillo,Gustavo.isaza}@ucaldas.edu.co
[2] Centro Nacional de Investigación del Café – CENICAFÉ
{narmer.galeano,alvaro.gaitan}@cafedecolombia.com
[3] Universidad de Manizales
Depatamento de Ciencias e Ingeniería
lbertel@umanizales.edu.co

Abstract. The protein sequence analysis can deal with various approaches in order to find the phenotypic and functional characteristics of the gene structure. Fortunately there are many models of genes already described in database with biological information.

Linking coffee gene annotations transcriptome to centralized biology information systems as Bio2RDF, offers the possibility of finding associated transcriptome relationships between them and between terms and concepts defined by semantic rules defined by ontologies.

Sesame was used as a repository to store information related triplets with coffee and transcriptomes index obtained from the Protein Data Bank (PDB), these relationships are the foundation for semantic search using SPARQL.

Data from the functional relationships searches are deployed through the endpoint provided by the repository from Sesame and Pubby.

Keywords: functional gene networks, Resource Description Framework RDF, Linked Data, Bio2RDF.

1 Introduction

Coffee is the most important agricultural product for Colombia and occupies the second line of importance in their economy after oil [1]. There are two principal species of coffee grown in Colombia: C. arabica and C. canephora ; the former is the source of mild coffee and the second is the producer of robusta coffee. These species are not unique, as over a hundred of them can be found in the wild, almost all in Africa. Using Bioinformatics in the acquisition and processing of coffee genetic data, has provided valuable information of different types of coffee species. Genomic characterization has provided the ability to identify desirable qualities such as resistance to pests and diseases, environmental adaptation, bean yield per plant and

M.S. Mohamad et al. (Eds.): 7th International Conference on PACBB, AISC 222, pp. 45–51.
DOI: 10.1007/978-3-319-00578-2_7 © Springer International Publishing Switzerland 2013

flavor characteristics, important to keep Colombia to the forefront of the coffee production at world.

Bioinformatics tools such as BLAST and Interproscan [2] were used to the structural construction of model genes based on sequence data, these tools were also used to identify the molecular structures that allow defining the function of genes and proteins. Using these bioinformatics tools marks the first stage of a macro process, building networks between genes comprehensively[3], that could explain the biochemical processes of cells and then give the phenotype that characterizes each organism.

Mining techniques using the information contained in the web pages so the markup language for web pages (HTML) is not suitable for processing, to fix this problem nowadays formats that allow the automation of search and exchange of information are used, usually deployed in extensible markup language format (XML). The structure format given by XML has the advantage of recognizing the type of information described[4].

Although the use of XML is an important step for the exchange of information, it does not provide the necessary requirements to implement semantic search; instead, languages such as Web Ontology Language (OWL) and the Resource Description Framework (Resource Description Framework, RDF) are used. Especially RDF allows the definition of relationships of Internet resources (web pages, images, URL, multimedia, etc.), and it is used to target available resources in the Internet. When connecting these resources with RDF, graphs are created that interconnect concepts, and can be used to search relationships.

Initiatives such as Linked Data have permitted to create graphs of relations in various fields of knowledge [5], Figure 1. For the field of bioinformatics, the project Bio2RDF brings together biological information generated by more than one hundred database used in the investigation of the life sciences. Linking data of coffee genetic material annotations with the resources outlined in Bio2RDF, creates a graph that allows the search for functional relationships.

2   Bio2RDF

Bio2RDF (http://bio2rdf.org) gathers information from various sources related to bioinformatics, Figure 1, using Semantic Web technology, based on RDF, OWL and SPARQL. The entire information contained in Bio2RDF exposed as RDF resources.

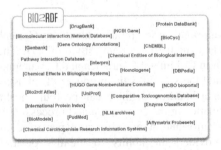

Fig. 1. Domains of biological information contained in Bio2RDF

The databases containing biological data are classified by namespace, which are unique for each database. If you wish to make an inquiry to Gene Ontology, would use the namespace *go* followed by the value of the biological component. To check the ID of the 15275 gene in the NCBI data set; would be: http://bio2rdf.org/geneid:15275. Table 1 is a compilation of namespaces and RDF triplets associated to databases contained in Bio2RDF.

Table 1. Some Namespace used in Bio2RDF (updated in 2012)

Dataset	Number of RDF	Namespace
NCBI Gene	394026267	geneid:
Gene Ontology	80028873	go:
Integrated resource of protein families, domains and functional sites [interpro]	999031	ipr:
Kyoto Encyclopedia of Genes and Genomes [kegg]. Compuesta de 16 bases de datos.	49850774	cpd: (compount)
		dr: (drug)
		ec: (enzyme)
		gl: (glycan)
		rn: reaction
		path: pathway
Protein Data Bank		pdb:

The Bio2RDF information search is done through SPARQL language, queries can be generated locally or remotely, the results are usually returned in RDF format. Bio2RDF architecture provides endpoint for users to perform remote queries. To access through web to the Bio2RDF query manager the URL http://atlas.bio2rdf.org/sparql should be accessed. If for example, you want to know the publications in pubmed for the term DEAMINASE, the query would be:

```
select distinct ?g ?o
where {
graph ?g
{
?s ?p ?o .
?o bif:contains "DEAMINASE" .
}
}
```

A summary of the results is shown in Table 2.

External data of biological data can be obtained in two ways: the available data exchange format based on XML and direct access to databases such as Entrez Gene, GO, PDB, etc.. These data are built and stored in a repository that can be Sesame or Virtuoso OpenLink. The deployment of information can be done through website via PUBBY or through web services as REST or SOAP [6].

Table 2. Summary of the result of the query with SPARQL

G	o
http://bio2rdf.org/pubmed	Ancestral founder of mutation W283X in the porphobilinogen deaminase gene among acute intermittent porphyria patients. [pubmed:12566739]
http://bio2rdf.org/pubmed	De novo protein synthesis is required for the activation-induced cytidine deaminase function in class-switch recombination. [pubmed:12591955]
http://bio2rdf.org/pubmed	Coordinate induction of AMP deaminase in human atrium with mitochondrial DNA deletion. [pubmed:12604357]
http://bio2rdf.org/pubmed	E-proteins directly regulate expression of activation-induced deaminase in mature B cells. [pubmed:12717431]

3   Materials and Methods

Coffee sequencing infromation was generated by the National Coffee Research Center of Colombia (Cenicafe). The data are obtained after making a filtering process and verification with bioinformatics tools such as BLAST and InterProScan. With the annotations obtained biological information is recovered through the search application in databases such as Protein Data Bank (PDB), Gene Ontology, etc.. The results are associated by means of RDF to the information found in Bio2RDF; this process is known as data reification. CEN notation is used to denote the provided annotations by CENICAFE, for example, CEN436416 is associated to PDB data using the identifier for Bio2RDF as http://bio2rdf.org/pdb:1tdj_A, also defines a biochemical concept associated with the annotation, for our example is DEAMINASE.

For data annotation: Annotacion{cen = 'CEN436416' tdj_A pdb = '1 ', description =' Deaminase '}; RDF content is:

```
<rdf:RDF
    xmlns:rdf="http://www.w3.org/1999/02/22-rdf-syntax-ns#"
xmlns:dc="http://purl.org/dc/terms/"
    xmlns:link="http://bio2rdf.org/bio2rdf_resource:"
    xmlns:recurso="http://bio2rdf.org/pdb:"
    xmlns:cenicafe="http://cenicafe.org/anotacion/cen/"
    xmlns:rdfs="http://www.w3.org/2000/01/rdf-schema#" >
    <rdf:Description
rdf:about="http://cenicafe.org/anotacion/cen/CEN436416">
        <link:linkedToFrom rdf:resource="http://bio2rdf.org/pdb:1tdj_A"/>
        <rdfs:label>DEAMINASE</rdfs:label>
        <dc:identified>CEN436416</dc:identified>
        <dc:title>CEN436416</dc:title>
    </rdf:Description>
  </rdf:RDF>
```

Figure 2 illustrates the process of linking data annotations to resources provided in Bio2RDF.

The starting points are the annotations which are then applied to bioinformatic search tools such as Gene Ontology, Interproscan, PDB, etc. The results provide information for linking to Bio2RDF through RDF format.

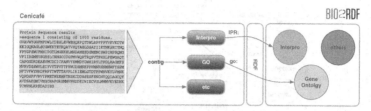

Fig. 2. Schematic of the annotation data reification and its relationship to Bio2RDF

With the linking of data annotations to Bio2RDF, a graph is created; containing annotations as starting point, the paths of the graph are provided by the information associated with Bio2RDF. Finding relationships between annotations of coffee genetic material, through the resources outlined in the graph is the key objective for this research. To find, for example, the annotations that have the id PDB associated to the term RIBOSOMAL. The result could indicate functional relationships between various annotations coffee genetic material.

The reifications were stored in the repository Sesame [7] with visualization through web pages by Pubby.

4   Results and Discussion

The use of graphs formed by the resource associations of coffee genetic material annotations with Bio2RDF, provides the ability to find relationships across the path of the graph. SPARQL query language [8] was used to query the relationship . The total number of triplets stored in the repository is 1650.

The first query performed was the search for CEN annotations that were associated in the same PDB.

This query found CEN396594 associated annotations, the association here were PDB. The results are illustrated in Figure 3.

Cen	Label	Pdb
cenicafe:CEN396594	"HYDROXYMETHYLTRANSFERASE"	<http://bio2rdf.org/pdb:1bj4_A>
cenicafe:CEN398177	"HYDROXYMETHYLTRANSFERASE"	<http://bio2rdf.org/pdb:1bj4_A>
cenicafe:CEN399089	"HYDROXYMETHYLTRANSFERASE"	<http://bio2rdf.org/pdb:1bj4_A>
cenicafe:CEN399090	"HYDROXYMETHYLTRANSFERASE"	<http://bio2rdf.org/pdb:1bj4_A>
cenicafe:CEN402266	"HYDROXYMETHYLTRANSFERASE"	<http://bio2rdf.org/pdb:1bj4_A>
cenicafe:CEN405067	"HYDROXYMETHYLTRANSFERASE"	<http://bio2rdf.org/pdb:1bj4_A>
cenicafe:CEN405058	"HYDROXYMETHYLTRANSFERASE"	<http://bio2rdf.org/pdb:1bj4_A>
cenicafe:CEN410664	"HYDROXYMETHYLTRANSFERASE"	<http://bio2rdf.org/pdb:1bj4_A>
cenicafe:CEN422573	"HYDROXYMETHYLTRANSFERASE"	<http://bio2rdf.org/pdb:1bj4_A>
cenicafe:CEN424631	"HYDROXYMETHYLTRANSFERASE"	<http://bio2rdf.org/pdb:1bj4_A>

Fig. 3. Query result by association of PDB

Figure 4 illustrates the results of the search of the scores related to coffee genomics by means of the ratio of the proportion-tuple information for PDB. The search entry is represented with orange oval (CEN396594), association with PDB is performed through Bio2RDF (1bj4_A) associated annotations are represented by blue ovals.

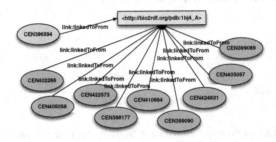

Fig. 4. Results of the functional relationships associated PDB

5   Conclusion and Future Work

Using tools like Linked Data and particularly Bio2RDF allowed the data linking of coffee gene annotations to shared resources and especially the Protein Data Bank (PDB) set.

The search for functional relationships of genes can be done at a first level by associations that are achieved through the graph formed between annotations and Bio2RDF, for this case the associations between annotations and the PDB and these annotations and biochemical component labels are used.

The next step will be to expand the search for relationships using the terms defined in the description of the information in the Protein Data Bank. Another feature to explore is the application of inference using ontologies defined in Bio2RDF, especially the equivalence of concepts as used in OWL.

It is important to deploy the relationships in a component that provides graphical information of the different connections that are founded in the searches.

References

[1] Hartwick, E.: The Cultural Turn in Geography: A New Link in the Commodity Chain. In: Encounters and Engagements between Economic and Cultural Geography. GeoJournal Library, vol. 104, pp. 39–46 (2012)

[2] Li, W., Feng, J., Jiang, T.: Workshop: Transcriptome assembly from RNA-Seq data: Objectives, algorithms and challenges. In: IEEE 1st International Conference on Computational Advances in Bio and Medical Sciences (ICCABS), p. 271 (2011)

[3] Castillo, L.F., Galeano, N., Isaza, G.A., Gaitan, A.: Construction of coffee transcriptoma networks based on gene annotation semantics. Journal of Integrative Bioinformatics, 1–13 (2012)

[4] Coulet, A., Shan, N.H., Garten, Y., Musen, M., Altman, R.B.: Using text to build semantic networks for pharmacogenomics. J. Biomed. Inform. 43(6), 1009–1019 (2010)

[5] Heath, T., Bizer, C.: Linked Data: Evolving the web into global Data Space. Morgan & Claypool Publishers series (2011)

[6] Belleau, F., Nolin, M.-A., Tourigny, N., Rigault, P., Morissette, J.: Bio2RDF: Towards a mashup to build bioinformatics knowledge systems. Journal of Biomedical Informatics 41, 706–716 (2008)

[7] Broekstra, J., Kampman, A., Van Harmelen, F.: Sesame: An Arquitecture for Storing and Querying RDF Data and Schema Information. Spinning the Semantic Web, 197–221 (2005)

[8] DuCharme, B.: Learning SPARQL: Quering and Updating with SPARQL 1.1 (2011)

A Cellular Automaton Model of the Effects of Maspin on Cell Migration

M.A. Al-Mamun[1], M.A. Hossain[1], M.S. Alam[2], and R. Bass[1,3]

[1] Computational Intelligence Group, Faculty of Engineering and Environment,
University of Northumbria at Newcastle, UK
{mohammed.al-mamun,alamgir.hossain}@northumbria.ac.uk
[2] Department of Applied Physics, Electronics and Communication Engineering,
University of Dhaka, Dhaka-1000, Bangladesh
msalam@univdhaka.edu
[3] Department of Applied Sciences, Faculty of Health and Life Sciences,
University of Northumbria at Newcastle, UK
rosemary.bass@northumbria.ac.uk

Abstract. Maspin (Mammary Serine Protease Inhibitor) is a non-inhibitory serpin with multiple cellular effects that is a type II tumour metastasis suppressor. Maspin has been shown to reduce cell migration, invasion, proliferation and angiogenesis, and increase apoptosis and adhesion. In this paper, we report the development of a mathematical model of the effects of maspin on cellular proliferation and migration. An artificial neural network has been used to model the unknown cell signalling to determine the cells fate. Results show that maspin reduces migration by between 10-35%; confirmed by published *in vitro* data. From our knowledge, this is the first attempt to model maspin effects in a computational model to verify *in vitro* data. This will provide new insights into to the tumour suppressive properties of maspin and inform the development of novel cancer therapy.

Keywords: Maspin, Serpin, Cell migration, Mathematical model, Neural Network.

1 Introduction

Maspin (SERPINB5) is a member of the serine protease inhibitor (serpin) superfamily which has been characterized as a type II tumour metastasis suppressor in multiple cancer types. Metastasis is a complex and multi-step process involving cell migration, invasion through the lamina propria, and growth in an extraneous microenvironment. Maspin decreases tumour growth and metastasis in vivo [1] and invasion in vitro [2]. This is achieved by the ability of maspin to influence aspects of cell behaviour including migration, invasion, proliferation, angiogenesis and apoptosis. These effects are proposed in many in vitro and in vivo models to involve both intracellular and extracellular activities of maspin. This diversity motivates us to build a computational model of the effects of maspin to show its potential engagement with multiple cellular phenomena using cellular automata modelling techniques.

M.S. Mohamad et al. (Eds.): *7th International Conference on PACBB*, AISC 222, pp. 53–60.
DOI: 10.1007/978-3-319-00578-2_8   © Springer International Publishing Switzerland 2013

The cellular mechanisms that maspin uses to influence cellular behaviour are not yet clearly defined, but have been reviewed recently [3]. There are reports that maspin works inside and outside the cell. It is possible that extracellular maspin directly affects cell migration, adhesion and angiogenesis, while indirectly affecting tumour cell proliferation and apoptosis. Maspin has been reported to bind to integrin cell adhesion receptors [4] or to the extracellular matrix [5]. Intracellular maspin binding partners have been identified, including the transcription factor IRF-6[2, 6], and histone deacetylase 1 influencing the Bcl-2/Bax signal axis [7, 8]. To date there have been no reports of mathematical and computational models to support these data. In this paper we have taken the migration raw data from [9], where it was reported that the G-helix is essential and sufficient for the influence of maspin on cell migration.

This paper presents an investigation into the development of a mathematical model of the effects of maspin on cellular proliferation and migration. The main objective of this investigation is to model proliferation and migration to investigate the unknown effects of maspin; this will allow us to verify the in vivo or in vitro experiments. The proposed model was implemented, tested and verified through a set of experiments to demonstrate the merits and capabilities of the scheme.

2   The Model

Mathematical models have been developed for tumour growth, tumour invasion and considerations of the different stages of tumour pathogenesis. Much work has been done to model the different aspects of tumour development. The majority of modelling approaches have considered all cancer cells as having the same properties. They considered the whole tumour mass as a single entity and defined the global parameters for every cell [10]. There are some attempts to model tumour growth characteristics at the cellular level as well, but considered cells as static entities. A cell is a complex living structure and its behaviour is not completely understood, especially when considering what has done wrong to allow the development of cancer. It evolves during both the growth process and therapy. A model discussing the evolutionary aspects of tumours at the cellular level has been presented [11]. Author considered the cell as an individual entity and modelled its decisions making processes based on the tumour microenvironment using a neural network. The model explored the consumption of oxygen, glucose concentration and hydrogen ions production but passed over some important growth constraints. In our previous model we established an in silico model to calculate tumour mass with the consideration of oxygen, glucose, extracellular matrix (ECM), cell-cell adhesion and cell movement as key micro environmental parameters. We also integrated the information regarding protein expression, growth promoters and growth inhibitors as tumour growth constraints and a bioreductive drug (TPZ) transport model for the hypoxic tumour cells [12]. In this paper we modified our neural network to show the effect of maspin presence/absence on tumour growth in respect of proliferation and migration [12].

A tumour tissue was developed starting with one cell at the centre of a two dimensional grid. In this cellular automaton model, each grid element was either occupied by an abnormal cell or was empty. The grid elements had local concentrations of oxygen, glucose and hydrogen ions. Each tumour cell was treated as an individual entity and identified through its position coordinates x(xx, yy). The cell was strongly influenced by tumour microenvironments and has its own decision mechanism that determined its behaviour during the growth activity. In this investigation, the decision mechanism was implemented using an artificial neural network. The neural network got the number of neighbours, the concentration of oxygen, the glucose concentration and the hydrogen ions as inputs. The response of the network at its output nodes determined the decision or behaviour of the cell. The response was one of three cellular behaviours: proliferation, quiescence and movement.

We consider the normal tumour growth with oxygen supply mentioned in [11]. As we have not considered the hypoglycemia (glucose effect) and acidity (hydrogen effect) in case of maspin during the in vitro experiments, we limited our model only in oxygen nutrient for the normal growth model. During the maspin model the oxygen concentration is taken constant as the oxygen and other nutrients were constant throughout the experiment. The varying distances of the cells from the blood vessel can cause heterogeneity in the tumour microenvironment and the tumour mass. The evolution of oxygen with respect to time was maintained by the following PDE defined [11].

$$\frac{\partial O_2(x,t)}{\partial t} = D_{O_2}\Delta O_2(x,t) - f_{O_2}(x,t) \tag{1}$$

In case of maspin, we formed the new diffusion equation for maspin in the model

$$\frac{\partial M(x,t)}{\partial t} = D_M\Delta M(x,t) - f_M(x,t) \tag{2}$$

The term $f_l(x, t)$ was the utilization or production function of oxygen, glucose and hydrogen ions ($l=O2$) for each cell at specific position x and at time t and is described in equation 3.2,

$$f_l(x,t) = \begin{cases} 0 \rightarrow & \text{number of tumour cell at that grid point} \\ cr_i \rightarrow & \text{grid point was located by the tumour cell} \end{cases} \tag{3}$$

Where, cr_i are the base consumption/production rates $i=O_2$ and M, F(x) is the modulated energy consumed by the cell located at the grid element x and calculated in equation 3.3. It was used to report the differences for the energy consumptions among different subclones.

$$F(x) = \max(k(R - T_r) + 1, 0.25) \tag{4}$$

Where k is the strength of modulation, R is the response of the neural network and Tr is the target response. The term max (, 0.25) shows that the cell's metabolism was at least a quarter of the base line consumption rate as considered by Anderson [8]. This function also ensures that the cell with the greatest network response will consume more nutrients.

The tumour cell density $no= 0.0025-2 = 1.6 \times 105$ cells cm-2 because the cell resides on a 2D grid. All presented models discretized partial differential equations using standard five-point finite central difference formulas and used length scale $\Delta d = 0.0025$ (because in [13], it is reported the size of a real cancer cell is approximately equal to 25 μm) and time scale $\Delta t = 5 \times 10$-4. The parameter values incorporated during the tumour development process are given in table 1.

Table 1. Parameters used during tumour growth

Value	Definition	Value	References
D_{O2}	Oxygen diffusion constant	$1.8 \times 10^{-5} \text{cm}^2\text{s}^{-1}$	Grote et al. (1997) [14]
D_M	Maspin diffusion constant	$1 \times 10^{-6} \text{cm}^2\text{s}^{-1}$	Young et al. [15]
n_o	Cancer cell density	$1.6 \times 10^5 \text{cellscm}^{-2}$	Casciari et al. [13]
cr_{O2}	Base oxygen consumption rate	$2.3 \times 10^{-16} \text{mol}$ $\text{cells}^{-1}\text{s}^{-1}$	Freyer and Sutherland (1986) [16]
k	Modulation strength	6	Model specific
T_r	Target response	0.675	Model specific

The neural network structure developed here for the tumour cell growth model was divided into three layers: an input layer I, a hidden layer H and an output layer O. The number of specific cell's neighbours n(x, t) and its local chemical concentrations; oxygen O2(x, t) was fed as inputs into the network. The connections between the input and the hidden layer w, and then the hidden and the output layer W were established using some weighted matrices. The network weights and thresholds are chosen only to produce the correct output response. Standard tan-sigmoid is used a propagation function for the hidden and output layers. The nodes at the output layer were divided into proliferation, quiescent and movement phenotypes. The model checked the values at the first three output nodes and selected the greatest one as the life-cycle phenotype for the cell. The cell activated its movement when the movement node got a value greater than 0.5, a situation where available oxygen was reduced. In case of maspin similar neural network except the oxygen input is replaced with maspin and oxygen supply was considered as constant. The weights and biases were chosen by an iterative process to get the desired neural network response.

The fig.1 shows the basic structure of the neural network that models the behaviour of each tumour cell due to the impact of the tumour microenvironment. As in experiments we haven't taken any data for tumour apoptosis we haven't considered the apoptosis at the output of the neural network. It suggests that our model is only run for 13 hours duration which shows the migration of two kinds of tumour cell (proliferating and quiescent cells). All other model facts are remained same as [12].

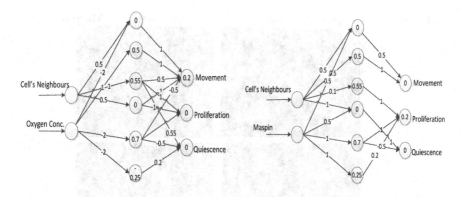

Fig. 1. The basic structure of NN for normal tumour growth (left) and Maspin tumour growth (right)

3   Results and Discussion

The initial simulation started from one cell at the centre of a grid, the cell divided and placed its daughter cell in one of the adjacent neighbourhoods after meeting some standards. At every new time step each cell performed one of its lifecycle actions; proliferation, quiescence and movement by selecting the highest neural network response and involved the production/consumption of chemical (oxygen). The chemical equations (1) and (2) are discretized using five –point finite central difference formulas with space step d and time step Δt. The cell consumed nutrients according to the selected phenotype; proliferating cells rapidly divide into daughter cells so they consume more nutrients than the stationary ones. The daughter cells inherited attributes from their parents and generated their behaviour by using these attributes as an input vector for the neural network. In every time step cells are updated in a random order. The model was simulated with oxygen concentrations as the determining factors of the cell behaviour. In the maspin model, the cell consumed maspin according to the selected phenotype of neural network output. The maspin is consumed as a form of diffusion and the cell took the maspin as same process mentioned above for the normal tumour growth model.

The fig 2 shows the migration of tumour cells with differential maspin expression during tumour growth in a visual form. Here we presented the tumour cell growth for different values of t up to 100. From the figure it can be seen that as time increases the presence of maspin reduces the cell migration, this means that the tumour cell density increases in comparison to control or normal growth of tumour. In brief cell movement is reduced. In our model, we can see that the invasion distance was decreased for tumour cells expressing maspin. Experimentally maspin reduces cell migration by up to 30 – 60% using different cancer cell lines in vitro [9]. We calculated the percentage of reduction in cell migration as being up to 10-35%, this supports the migration hypothesis made in [9].

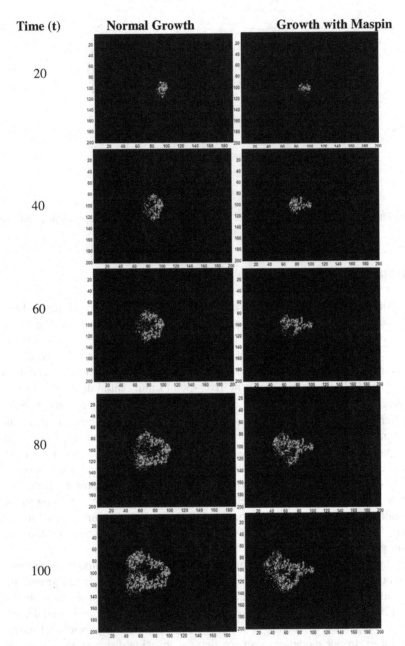

Fig. 2. Cell migration comparison between normal growth and growth in the presence of maspin at different time t=20, 40, 60...100

Maspin reduces cell proliferation and fig.3 shows the growth rate for both models. The cell counting shows that maspin reduces the proliferation in the model. Our next target will be to verify the cell proliferation rate with in vitro data.

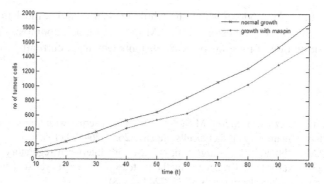

Fig. 3. Tumour growth rate during the model

To verifying the experimental data we have plotted migration data as a percentage for the both models. We have taken 6 samples from the model data, which shows that maspin reduces migration by 10-35% in comparison with normal growth. This fits with the data from the experiments; the range of migration reduction was determined experimentally to be 30-60% for different cell lines [9]. The samples have been taken from the model using the same techniques as in the cell migration assay *in vitro* experiments. The cell paths for 100 cells were chosen randomly selected cells. Because of this there are variations in different data sample which mimics the exact cell migration assay [9]. This indicates that our model can verify the cell migration effects due to maspin. More details about experimental approach will be found in [9].

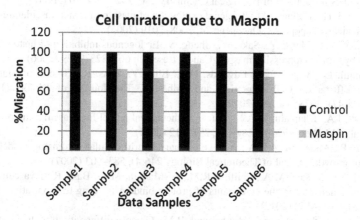

Fig. 4. Maspin effects on cell migration mathematical model

4   Conclusion

This paper has presented a cellular automaton model to show the proliferation and migration effect on tumour cells due to maspin. A mathematical model has been implemented, tested and verified using in-vitro data reported earlier [9]. Results demonstrate that the proposed model resemble the in-vitro experiment. Our future works will be to establish the model to show the whole unclear complex mechanism of maspin in respect of cell-cell adhesion, cell-ECM interaction and apoptosis. This will be informed by definition of maspin specific constants which are currently unavailable.

References

1. Zou, Z., Anisowicz, A., Hendrix, M.J., et al.: Maspin, a serpin with tumor-suppressing activity in human mammary epithelialcells. Science 263, 526–529 (1994)
2. Bailey, C.M., Abbott, D.E., Margaryan, N.V., et al.: Interferon regulatory factor 6 promotes cell cycle arrest and isregulated by the proteasome in a cell cycle-dependent manner. Molecular and Cellular Biology 28, 2235–2243 (2008)
3. Bodenstine, T.M., Seftor, R.E., Khalkhali-Ellis, Z., Seftor, E.A., Pemberton, P.A., Hendrix, M.J.: Maspin: molecular mechanisms and therapeutic implications. Cancer Metastasis Rev. 31(3-4), 529–551 (2012)
4. Cella, N., Contreras, A., Latha, K., Rosen, J.M., Zhang, M.: Maspin is physically associated with $\beta 1$ integrin regulating cell adhesion in mammary epithelial cells. FASEB Journal 20, 1510–1512 (2006)
5. Blacque, O.E., Worrall, D.M.: Evidence for a direct interaction between the tumor suppressor serpin, maspin, and types I and III collagen. Journal of Biological Chemistry 277, 10783–10788 (2002)
6. Bailey, C.M., Khalkhali-Ellis, Z., Kondo, S., Margaryan, N.V.V., Seftor, R.E.B., Wheaton, W.W., Amir, S., Pins, M.R., Schutte, B.C., Hendrix, M.J.: Mammary serine protease inhibitor (Maspin) binds directly to interferon regulatory factor 6: identification of a novel serpin partnership. Journal of Biological Chemistry 280, 34210–34210 (2005)
7. Li, Z., Shi, H.Y., Zhang, M.: Targeted expression of maspin in tumor vasculatures induces endothelial cell apoptosis. Oncogene 24, 2008–2019 (2005)
8. Li, X., Yin, S., Meng, Y., Sakr, W., Sheng, S.: Endogenous inhibition of histone deacetylase 1 by tumor-suppressive maspin. Cancer Research 66, 9323–9329 (2006)
9. Ravenhill, L., Wagstaff, L., Edwards, D.R., Ellis, V., Bass, R.: G-helix of Maspin Mediates Effects on Cell Migration and Adhesion. The Journal of Biological Chemistry 285(47), 36285–36292 (2010)
10. Sherratt, J.A., Chaplain, M.A.J.: A new mathematical model for avascular tumour growth. Journal of Mathematical Biology 43, 291–312 (2001)
11. Gerlee, P., Anderson, A.R.A.: An evolutionary hybrid cellular automaton model of solid tumour growth. Journal of Theoretical Biology 246(4), 583–603 (2007)
12. Kazmi, N., Hossain, M.A., Phillips, R.M., Al-Mamun, M.A., Bass, R.: Avascular tumour growth dynamics and the constraints of protein binding for drug transportation. J. Theor. Biol. 313, 142–152 (2012)
13. Casciari, J.J., Sotirchos, S.V., Sutherland, R.M.: Glucose diffusivity in multicellular tumor spheroids. Cancer Research 48, 3905–3909 (1988)
14. Grote, J., Susskind, S., Vaupel, P.: Oxygen diffusivity in tumour tissue (DS-carcinosarcoma) under temperature conditions within the range of 20-40 °C. Pflugers Architecture 372, 37–42 (1997)
15. Young, M.E., Carroad, P.A., Bell, R.L.: Estimation of diffusion coefficients of proteins. Biotechnol. Bioeng. 22(5), 947–955 (1980)
16. Mueller-Klieser, W., Freyer, J.P., Sutherland, R.M.: Influence of glucose and oxygen supply conditions on the oxygenation of multicellular spheroids. British Journal of Cancer 53, 345–353 (1986)

Quantitative Characterization of Protein Networks of the Oral Cavity

Fernanda Correia Barbosa[1,3], Joel P. Arrais[2], and José Luís Oliveira[1]

[1] Department of Electronics, Telecommunications and Informatics (DETI), Institute of Electronics and Telematics Engineering of Aveiro (IEETA), University of Aveiro, Portugal
[2] Department of Informatics Engineering (DEI), Centre for Informatics and Systems of the University of Coimbra (CISUC), University of Coimbra, Portugal
[3] Department of Informatics and Systems Engineering (DEIS), Higher Institute of Engineering of Coimbra (ISEC), Polytechnic Institute of Coimbra, Portugal

Abstract. Modeling protein interactions as complex networks allow applying graph theory to help understanding their topology, to validate previous evidences and to uncover new biological associations. Topological properties have been recognized by their contribution for the understanding of the structures, functional relationships and evolution of complex networks, helping in a better comprehension of the diseases mechanisms and in the identification of drug targets. The human interactome, i.e. the network formed by all protein-protein interactions, is a complex and yet unknown system.

In this paper we present the results of a study about the topological properties of the oral protein network. We evaluate several confidence scores and prediction methods, in order to compare these networks with random organizations with the same size.

Keywords: human oral proteome, complex networks, protein-protein interactions.

1 Introduction

Interactions between the elements of a cell, in particular protein-protein interactions (PPI) are responsible for the biological functions of the living species. The model of these interactions is useful to help understanding the functional relationships obtained through several techniques, often with high error rates, in different biological contexts. This proteomic studies are manually performed by researchers and require highly specialized knowledge.

PPI can be modeled as complex networks, where proteins are represented by nodes, and interactions represented by edges. Biological networks are generally sparsely connected, which is considered an evolutionary advantage for preserving robustness to random failures, and tend to be heterogeneous, with few nodes highly connected (hubs) and many nodes with few connections [1-4,5]. The study of the topological properties of complex networks allows understanding their structures and highlighting some similarities, like small world properties [6], power-law degree

M.S. Mohamad et al. (Eds.): *7th International Conference on PACBB*, AISC 222, pp. 61–68.
DOI: 10.1007/978-3-319-00578-2_9 © Springer International Publishing Switzerland 2013

distributions to distinguish from random and non-random networks [2], high average clustering coefficient (C_{avg}) showing modularity [5,7-9], and clustering degree distribution to identify hierarchies in their organization [2]. The degree distribution of most of the biological networks approximates a power-law, so they are named scale-free [1-3,4] and in most cases only the tail of the distribution follows a power-law, existing a value x_{min} for which the power-law is observed [3,4]. However, this is still a controversial subject as some researchers argue that PPI networks do not necessarily follow a power-law [10].

Several methods can be applied to detect and characterize power-law distributions. A simple one is the least-squares fitting of the distribution by a straight line in a log-log plot. More accurate methods are the maximum-likelihood fitting methods with goodness-of-fit tests based on the Kolmogorov-Smirnov (KS) statistic that give the slope of the fitted line, the value of x_{min} and a p-value, where is considered that for $p \leq 0.1$ the power-law model must be ruled out [3,4].

Several researchers evaluated the topological properties of different kinds of networks: Newman [1,3] evaluated the topological properties of twenty-seven datasets from different areas, like social, biological and technological; Colliza et al. [8] evaluated the topological properties of three distinct PPI networks of *S. cerevisiae*; Liu et al. [11] evaluated the topological properties of classical music from Bach, Mozart, Chopin, and Chinese pop music; Clauset et al. [4] evaluated twenty-four datasets from different areas, like physics, earth sciences, computer and information sciences, two of them being the degrees of proteins in the PPI network of the *yeast S. cerevisiae* and the degrees of metabolites in the metabolic network of the *bacterium E. coli*.

The human oral proteome is a subsystem of the human proteome complex system and their proteins were obtained from proteomic studies done by researchers. Here we propose to do an exploratory study of the human oral PPI networks, with different confidence scores, and from different prediction methods. This study includes the analysis of relevant topological properties of the human oral PPI network dataset, the comparison to respective random networks, the analysis of their degree distribution that is supposed to follow a power-law distribution and the evaluation of that assumption.

2   Methods

To better understand the organization and evolution of the human oral proteome, several networks are constructed for different confidence scores and for different prediction methods to model their PPI and various network topological measurements are used. The topological properties of the networks computed include the number of nodes and edges, the connected components, the network diameter, radius, density, centralization and heterogeneity, the C_{avg} and the corresponding average clustering coefficient for a random network (C_{rand}) [7-9], the characteristic path length, and the distributions of node degrees and clustering degrees [12,13]. Studied networks degree distributions are fitted to the power-law model and the corresponding p-value is measured using maximum-likelihood fitting methods with goodness-of-fit tests based on the KS statistic [4].

2.1   Dataset Construction

The human oral proteome dataset was obtained from proteomic studies done by researchers [14]. STRING (Search Tool for the Retrieval of Interacting Genes) is an online database resource [15] that, given several distinct types and sources of PPI information, aims to provide in one site an integration and evaluation service. Interactions in STRING are provided with a confidence score and are obtained from different prediction methods like Experiments, Co-occurrence, Co-expression, Databases, Neighborhood, Gene Fusion and Text Mining. STRING also presents additional information such as protein domains and 3D structures [16]. Using this dataset, several networks were constructed representing the entire set of PPI for different confidence scores ($> 100, > 200 \cdots > 900$) and for different prediction methods (Experiments, Co-occurrence, Co-expression, Databases and Neighborhood). These networks were constructed as undirected, unweight and with no self-edges.

2.2   Measurements of Network Topology

An undirected graph G can be defined as a pair $G = (V, E)$ where V is a set of vertices representing the nodes and E is a set of edges representing the connections between the nodes i and j. The number of nodes of a graph G is denoted by N and the number of edges of a graph is denoted by L. Given a graph $G = (V, E)$ the adjacency matrix representation consists of a $N \times N$ matrix $A = [a_{ij}]$, such that $a_{ij} = 1$ if $(i, j) \in E$ or $a_{ij} = 0$ otherwise. For undirected graphs the matrix is symmetric [1,5].

The average number of neighbors, denoted by $\langle k \rangle$, indicates the average connectivity of a node in the network, and the degree k_i of a node i in an undirected graph represents the number of neighbors of the node i [1,5,13,17].

The average clustering coefficient, C_{avg}, of the whole network, assuming that i is a vertex with degree k_i in an undirected graph G and that there are e_i edges between the k neighbors of i in G, is given by [1]

$$C_{avg} = \frac{1}{N}\sum_{i=1}^{N} C_i = \frac{1}{N}\sum_{i=1}^{N} \frac{2\sum_{l \neq i}\sum_{m \neq i,l} a_{il}a_{lm}a_{mi}}{\left(\sum_{l \neq i} a_{il}\right)^2 - \sum_{l \neq i} a_{il}^2} = \frac{1}{N}\sum_{i=1}^{N} \frac{2e_i}{k_i(k_i-1)} \tag{1}$$

with $0 \leq C_i \leq 1$ the local clustering coefficient of a node i in G. For random networks with the same properties of the considered datasets, C_{rand} is [7-9]

$$C_{rand} = \frac{1}{N}\frac{\left(\langle k^2 \rangle - \langle k \rangle\right)^2}{\langle k \rangle^3} \tag{2}$$

The network diameter, d, is the largest distance between two nodes. The average shortest path length or characteristic path length, gives the expected distance between two connected nodes. Eccentricity is the maximum non-infinite length of a shortest path between i and another node in the network. The maximum node eccentricity is the diameter. The network radius, r, is the minimum among the non-zero eccentricities of the nodes in the network. A normalized version of the average number of neighbors $\langle k \rangle$, is the density of a network, which varies between 0 and 1. Networks with a star-like topology have centralization close to 1, whereas decentralized networks are characterized by having centralization close to 0. The network heterogeneity reflects the tendency of a network to contain hub nodes [1].

If x represents the quantity whose distribution we are interested in, a continuous power-law distribution is described by a probability density $p(x)$ such that [3,4]

$$p(x)\,dx = P_r(x \leq X \leq x + dx) = Cx^{-\alpha}\,dx \tag{3}$$

where x is the observed value and C is a normalization constant. x_{min} is the lower bound for the power-law behavior. In the discrete case and in the case of integer values we have [3,4]

$$p(x) = P_r(X = x) = Cx^{-\alpha} \tag{4}$$

The fitting of power law forms to empirical distributions give some estimate of the slope α and the lower-bound x_{min}.

Another method of plotting the data is to calculate a cumulative distribution function (CDF), which also follows a power law, but with a different exponent $\alpha - 1$ [3]. The CDF of a power-law distributed variable $P(x)$ is, for the continuous and discrete cases, defined by [3,4]

$$P(x) = P_r(X \geq x) \tag{5}$$

Using the least-squares linear regression on the logarithm of the histogram to extract the slope α generates systematic errors [4]. The method of maximum likelihood for fitting power-law distributions to observed data gives accurate parameter estimates in the limit of large sample size [4]. If a distribution follows a power law exactly for $x \geq x_{min}$, the maximum likelihood estimator (MLEs) of the scaling parameter for the continuous case is [3,4]

$$\hat{\alpha} = 1 + n\left[\sum_{i=1}^{n} \ln \frac{x_i}{x_{min}}\right]^{-1} \tag{6}$$

where x_i, $i = 1 \cdots n$ are the observed values of i such that $x_i \geq x_{min}$. An approximate expression of $\hat{\alpha}$, if discrete power-law behavior is approximated by its continuous equivalent with x rounded to the nearest integer, is [3,4]

$$\hat{\alpha} = 1 + n\left[\sum_{i=1}^{n} \ln \frac{x_i}{x_{min}-\frac{1}{2}}\right]^{-1} \tag{7}$$

$\hat{x}_{min}$ is chosen to make the probability distributions of the measured data and the best-fit power-law model as similar as possible above $\hat{x}_{min}$ to both discrete and continuous data. It is used the KS statistic, which is the maximum distance between the

CDFs of the data and the fitted model. The estimate $\hat{x}_{min}$ is then the value of x_{min} that minimizes that distance. To quantify the uncertainty in the estimate of x_{min} it was used the "bootstrap" method [4].

Being roughly straight on a log-log plot is a necessary but not sufficient condition for power-law behavior. It is used a goodness-of-fit test, which generates a p-value that quantifies the plausibility of the power-law behavior and it is considered that the power law is ruled out if $p \leq 0.1$. To generate the synthetic data it is used the semi-parametric approach [4].

3   Results

Table 1 shows the topological properties measured for each studied PPI network: 1) number of proteins of the largest component (N); 2) % of proteins of the largest component regarding the total number of proteins for each confidence score or prediction method (%N (LC)); 3) number of interactions of the largest component (L); 4) % of interactions of the largest component regarding the total number of interactions for each confidence score or prediction method (%L (LC)); 5) C_{avg} (C_{avg}); 6) C_{rand} (C_{rand}); 7) diameter of the network (d); 8) radius of the network (r); 9) characteristic path length ($\langle d \rangle$); 10) average degree of the network ($\langle k \rangle$); 11) network density ($dens$); 12) network centralization ($cent$); and 13) network heterogeneity (h). The largest component represents almost the whole network in all networks except for the Co-occurrence network. Comparison between the C_{avg} of the studied networks with the correspondent C_{rand} indicates modularity (Figure 1) [2,7-9,17]. It was observed that with the increase of the confidence score the networks size, average degree, density and centralization decrease, but the diameter, radius, characteristic path length and heterogeneity increases. In the PPI networks from different prediction methods the diameter varies from 7 (Neighborhood) to 21 (Co-occurrence) and the characteristic path length varies from 2.80 (Neighborhood) to 7.31 (Co-occurrence).

Figure 2 shows the cumulative degree distribution of the studied networks. Basic parameters for the analysis of the degree distributions are calculated for the studied networks (Table 2): 1) size of the dataset (n); 2) average of the observed values ($\langle x \rangle$); 3) maximum of the observed values (x_{max}); 4) minimum of the observed values where the distribution follows a power-law (x_{min}); 5) error of x_{min} ($x_{min\,err}$); 6) slope of the fitted power-law (α); 7) slope error (α_{err}); 8) p-value (p).

For the PPI network of the yeast *Saccharomyces cerevisiae*, Clauset [4] obtained the value of $\alpha = 3.1 \pm 0.3$ and $p = 0.31$.

The studied datasets have values from $\alpha = 1.53$ for confidence score > 900 to $\alpha = 3.5$ for confidence score > 100 and > 200 and from $\alpha = 1.77$ for the Co-occurrence dataset to $\alpha = 2.57$ for the Neighborhood dataset. The p-values show that power-law distribution model is consistent for the networks with confidence score > 100, > 700 and > 800 and for all the networks of the prediction methods except the Co-occurrence network.

Table 1. Topological properties of the studied PPI networks with different confidence scores and different prediction methods

Description	N	%N (LC)	L	%L (LC)	C_{avg}	C_{rand}	d	r	<d>	<k>	dens	cent	h
CS>=0	3052	100.00%	200841	100.00%	0.302	0.001	6	3	2.23	131.61	0.04	0.27	0.91
CS>=100	3052	100.00%	200841	100.00%	0.302	0.001	6	3	2.23	131.61	0.04	0.27	0.91
CS>=200	3031	100.00%	151817	100.00%	0.275	0.001	6	3	2.33	100.18	0.03	0.23	0.94
CS>=300	2993	100.00%	90092	100.00%	0.237	0.003	6	4	2.59	60.20	0.02	0.13	0.90
CS>=400	2962	100.00%	61944	100.00%	0.237	0.013	7	4	2.84	41.83	0.01	0.01	0.96
CS>=500(LC)	2916	99.93%	47345	100.00%	0.257	0.024	7	4	3.07	31.47	0.01	0.01	1.04
CS>=600(LC)	2836	99.75%	35846	99.99%	0.282	0.042	9	5	3.35	25.28	0.01	0.09	1.11
CS>=700(LC)	2577	99.12%	26965	99.,94%	0.329	0.062	9	5	3.81	20.93	0.01	0.08	1.16
CS>=800(LC)	2359	98.74%	21204	99.91%	0.348	0.085	10	5	3.91	17.98	0.01	0.07	1.19
CS>=900(LC)	1840	93.59%	10944	98.84%	0.342	0.251	14	8	5.05	11.90	0.01	0.06	1.30
Experiments (LC)	2271	98.65%	17159	99.87%	0.183	0.136	9	5	3.61	15.11	0.01	0.15	1.49
Co-expression (LC)	1860	96.77%	51007	99.81%	0.467	0.002	13	7	3.49	54.85	0.03	0.18	1.28
Co-occurrence (LC)	393	53.47%	1668	70.23%	0.377	0.244	21	11	7.31	8.49	0.02	0.08	0.86
Databases (LC)	1341	91.22%	12573	98.26%	0.505	0.073	16	8	4.84	18.75	0.01	0.07	1.05
Neighborhood (LC)	466	99.57%	5030	99.98%	0.340	0.013	7	4	2.80	21.59	0.05	0.17	1.05

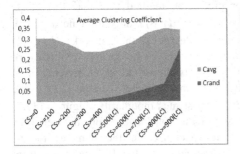
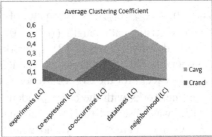

Fig. 1. C_{avg} of the studied PPI networks comparison with the corresponding C_{rand} with different confidence scores and different prediction methods

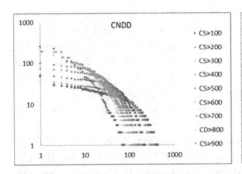
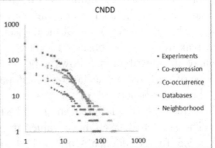

Fig. 2. Cumulative node degree distribution (CNDD) with logarithmic binning of the studied PPI networks with different confidence scores and different prediction methods

Clustering degree distributions are not independent of the degree, being decreasing with it (Figure 3). This gives evidence of some hierarchical modularity [2].

We can conclude that most of the studied networks generate scale-free networks with high degree of modularity and with some hierarchical organization. The small diameter also indicates small world properties.

Table 2. Basic parameters of the degree distributions of the studied PPI networks

Description	n	$<x>$	x_{max}	x_{min}	$x_{min\ err}$	α	α_{err}	p
>100	439	7.11	49	11	0.75	3.50	0.10	0.180
>200	352	8.61	48	15	4.04	3.50	0.57	0.006
>300	230	13.01	51	5	7.06	1.81	0.68	0.000
>400	175	16.93	73	4	8.02	1.69	0.53	0.000
>500 (LC)	156	18.69	101	17	10.38	2.17	0.51	0.000
>600 (LC)	133	21.32	159	7	12.05	1.76	0.45	0.004
>700 (LC)	117	22.03	192	15	7.94	2.05	0.32	0.298
>800 (LC)	103	22.90	231	12	6.32	1.95	0.25	0.350
>900 (LC)	76	24.24	255	2	8.97	1.53	0.30	0.056
Experiments (LC)	97	23.41	290	9	5.16	1.78	0.20	0.200
Co_expression (LC)	256	72.7	126	3	0.64	2.06	0.10	0.667
Co-occurrence (LC)	30	13.10	45	4	6.45	1.77	0.72	0.021
Databases (LC)	75	17.88	103	11	8.73	2.08	0.52	0.108
Neighborhood (LC)	75	12.21	42	8	2.48	2.57	0.47	0.601

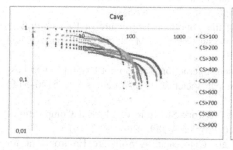 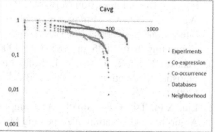

Fig. 3. Cumulative clustering degree distributions with logarithmic binning for the studied PPI networks with different confidence scores and different prediction methods

4   Conclusions

In this study, we evaluated the main topological properties of human oral PPI networks using different confidence scores and different prediction methods. The node degree distributions were fitted to the power-law model and the corresponding p-values were calculated, using maximum-likelihood fitting methods and goodness-of-fit tests based on the KS statistic.

Most of the studied networks generate scale-free networks with high degree of modularity and with some hierarchical organization and their small diameter indicates small world properties. While exploratory, this study aims to contribute to a better understanding of the human oral biology as a subsystem (less studied as his own) of the human biology system.

Acknowledgements. The research leading to these results has received funding from the European Community's Seventh Framework Programme (FP7/2007-2013) under grant agreement no. 200754 - the GEN2PHEN project.

References

1. Newman, M.E.J.: The Structure and Function of Complex Networks. SIAM Review 45, 167–256 (2003)
2. Barabási, A., Oltvai, Z.N.: Network Biology: Understanding the Cell's Functional Organization. Nature Reviews Genetics 5, 101–113 (2004)
3. Newman, M.E.J.: Power laws, Pareto distributions and Zipf's law. Contemporary Physics 46(5), 323–351 (2005)
4. Clauset, A., Shalizi, C.R., Newman, M.E.J.: Power-law Distributions in Empirical Data. SIAM Rev. 51(4), 661–703 (2009)
5. Pavlopoulos, G.A., Secrier, M., Moschopoulos, C.N., Soldatos, T.G., Kossida, S., Aerts, J., Schneider, R., Bagos, P.G.: Using Graph Theory to Analyze Biological Networks. BioData Mining 4, 10 (2011)
6. Watts, D.J., Strogatz, S.H.: Collective Dynamics of 'Small-World' Networks. Letters to Nature 393, 440–442 (1998)
7. Newman, M.E.J.: Random Graphs as Models of Networks. In: Bornholdt, S., Schuster, H.G. (eds.) Handbook of Graphs and Networks: from the Genome to the Internet, pp. 35–68. Wiley-VCH, Berlin (2003)
8. Colizza, V., Flammini, A., Maritan, A., Vespignani, A.: Characterization and Modeling of Protein-protein Interaction Networks. Physica A 352, 1–27 (2005)
9. Wu, D.D., Hu, X.: Mining and Analyzing the Topological Structure of Protein–Protein Interaction Networks. In: Proceedings of the 2006 ACM Symposium on Applied Computing, SAC 2006, pp. 185–189 (2006)
10. Tanaka, R., Yi, T.M., Doyle, J.: Some Protein Interaction Data do not Exhibit Power Law Statistics. Federation of European Biochemical Societies (FEBS) Letters 579, 5140–5144 (2005)
11. Liu, X.F., Tse, C.K., Small, M.: Complex Network Structure of Musical Compositions: Algorithmic Generation of Appealing Music. Physica A 389, 126–132 (2010)
12. Shannon, P., Markiel, A., Ozier, O., et al.: Cytoscape: A Software Environment for Integrated Models of Biomolecular Interaction Networks. Genome Research 13, 2498–2504 (2003)
13. Assenov, Y., Ramírez, F., Schelhorn, S.E., Lengauer, T., Albrecht, M.: Computing Topological Parameters of Biological Networks. Bioinformatics 24, 282–284 (2008)
14. Arrais, J.P., Rosa, N., Melo, J., Coelho, E.D., Amaral, D., Correia, M.J., Barros, M., Oliveira, J.L.: OralCard: A bioinformatic tool for the study of oral proteome. Archives of Oral Biology (2013) ISSN 0003-9969
15. Search Tool for the Retrieval of Interacting Genes, http://string-db.org
16. Szklarczyk, D., Franceschini, A., Kuhn, M., Simonovic, M., Roth, A., Minguez, P., Tobias Doerks, T., Stark, M., Muller, J., Bork, P., Lars, J., Jensen, L.J., Mering, C.: The STRING Da-tabase in 2011: Functional Interaction Networks of Proteins, Globally Integrated and Scored. Nucleic Acids Research 568, D561–D568 (2011)
17. Dong, J., Horvath, S.: Understanding Network Concepts in Modules. BMC Systems Biology 1 (2007)

Analysing Relevant Diseases from Iberian Tweets

Víctor M. Prieto[1], Sergio Matos[2], Manuel Álvarez[1],
Fidel Cacheda[1], and José Luís Oliveira[2]

[1] University of a Coruña, Department of Information and Communication Technologies,
Campus Elviña s/n, A Coruña, Spain
{victor.prieto,manuel.alvarez,fidel.cacheda}@udc.es
[2] University of Aveiro, DETI/IEETA, Campus Universitario de Santiago, 3810-193 Aveiro,
Portugal
{aleixomatos,jlo}@ua.pt

Abstract. The Internet constitutes a huge source of information that can be exploited by individuals in many different ways. With the increasing use of social networks and blogs, the Internet is now used not only as an information source but also to disseminate personal health information. In this paper we exploit the wealth of user-generated data, available through the micro-blogging service Twitter, to estimate and track the incidence of health conditions in society, specifically in Portugal and Spain. We present results for the acquisition of relevant tweets for a set of four different conditions (flu, depression, pregnancy and eating disorders) and for the binary classification of these tweets as relevant or not for each case. The results obtained, ranging in AUC from 0.7 to 0.87, are very promising and indicate that such approach provides a feasible solution for measuring and tracking the evolution of many health related aspects within the society.

Keywords: Data mining, classification, social media, detecting health conditions.

1 Introduction

The Internet constitutes a huge source of information that can be exploited for various needs. For a long time, it has been used by individuals seeking medical information. However, with the advent of the Web2.0 paradigm, the Internet is now used not only as an information source but also to disseminate personal health information, experiences and knowledge [12] [15].

Much of this health related information is shared through social media platforms such as Twitter and Facebook. Twitter[1], for example, offers a micro-blogging platform that allows users to communicate through status updates limited to 140 characters, commonly referred to as "tweets". It has over 200 million active users[2], and around 400 million tweets are published daily. These large quantities of user generated content (UGC) can been exploited in different ways and represent great opportunities for data and text mining approaches in many fields of application. Mining these data provides

---

[1] http://www.twitter.com
[2] https://twitter.com/twitter/status/281051652235087872

M.S. Mohamad et al. (Eds.): 7th International Conference on PACBB, AISC 222, pp. 69–76.
DOI: 10.1007/978-3-319-00578-2_10   © Springer International Publishing Switzerland 2013

an instantaneous snapshot of the public's opinions, and longitudinal tracking allows identification of changes in opinions [3]. This applies also to health related information, as can be verified by the various works that use Twitter and other user-generated data to assess and categorize the kind of information sought by individuals, to infer health status or measure the spread of a disease in a population. In [11], for example, the authors compared three web-based biosecurity intelligence systems and highlighted the value of social media, namely Twitter, in terms of the speed the information is passed and also because many issues or messages where not disseminated through other means. The greatest advantage of these methods over traditional ones is instant feedback: while health reports are published in a weekly or monthly basis, both tweets and query log of search engines can be obtained almost instantly. This characteristic is of extreme importance because early stage detection can reduce the impact of epidemic breakouts [1,7].

In this work, we propose an automated method, taking advantage of the wealth of data provided by Twitter, to measure the incidence of a set of health conditions in society, namely flu, depression, pregnancy and eating disorders. We focused our work on two official languages in the Iberian peninsula (Portuguese and Spanish), but the method we propose could be applied directly (e.g. South America) or adapted for other regions.

2 Related Work

Several works regarding the retrieval of health information from social media have already been published, with a major focus on measuring the incidence rate of influenza.

Chew and Eysenbach [3] suggested a complementary infoveillance approach during the 2009 H1N1 pandemic, using Twitter. They applied content and sentiment analysis to 2 million tweets containing the keywords "swine flu", "swineflu", or "H1N1". For this, they created a range of queries related to different content categories, and showed that the results of these queries correlated well with the results of manual coding, suggesting that near real-time content and sentiment analysis could be achieved, allowing monitoring large amounts of textual data over time. Signorini et al. [17] used Twitter to monitor public concern and levels of disease during the H1N1 pandemic in the United States. They collected tweets matching a set of 15 pre-specified search terms including "flu", "vaccine", "tamiflu", and "h1n1". They used content analysis to measure public interest and concern about this issue, and also applied support-vector regression to estimate influenza-like illness levels, using the Centers for Disease Control (CDC) data as reference. Using a model trained on 1 million influenza-related tweets, they reported average errors ranging from 0.04% to 0.93%. Lampos and Cristianini [10] and Culotta [5,6] also used regression models to estimate flu incidence rates in the United Kingdom and the United States respectively, obtaining correlation ratio of approximately 0.95. Aramaki et al. [1] applied SVM machine learning techniques to Twitter messages to predict influenza rates in Japan, achieving a correlation ratio of 0.89. Santos and Matos [14] combined data from Twitter and search engine logs in a regression model to estimate the incidence of flu in Portugal, achieving a correlation ratio of 0.89.

Chunara et al. [4] analysed cholera-related tweets published during the first 100 days of the 2010 Haitian cholera outbreak. For this, all tweets published in this period and containing the word "cholera" or the hashtag "#cholera" were considered, and these data were compared to data from two sources: HealthMap, an automated surveillance platform, and the Haitian Ministry of Public Health (MSPP). They showed good correlation between Twitter and HealthMap data, and showed a good correlation (0.83) between Twitter and MSPP data in the initial period of the outbreak, although this value decreased to 0.25 when the complete 100 days period was considered.

Apart from analysing the incidence of flu and infectious diseases related events, the analysis of other health parameters using Twitter data has also been reported. Scanfeld et al. [15], for example, applied content analysis to 1000 tweets to explore evidence of misunderstanding or misuse of antibiotics. Heaivilin et al. [9] also applied content analysis to a set of 1000 tweets matching search criteria relating to dental pain. The content was coded using pre-established categories, including the experience of dental pain, actions taken or contemplated in response to a toothache, impact on daily life, and advice sought from the Twitter community. Bosleya et al. [2] analysed and categorized 60 thousand tweets concerning cardiac arrest and resuscitation, obtained during a 38 day period using a set of 7 search terms. All these works have in common that the content analysis is performed manually. This fact limits their application over long time periods, as well as great amount of data or large regions.

3   Method

In this article we propose a method to extract a set of tweets that show the presence of certain health conditions in people, as a point of the departure to infer the incidence of such conditions in society. This section describes the procedures used to obtain the tweets related to these conditions.

In order to obtain the different sets of tweets, we defined several regular expressions to extract only the tweets related to the studied diseases.

To create these expressions, we initially obtained a set of tweets containing the name of each condition, removing re-tweets and tweets that included links, and calculated the log-likelihood of the words that occurred within those datasets, therefore obtaining an ordered list of words associated with each disease. Based on these lists, on manual content analysis and on general knowledge about the studied diseases, we then defined the regular expressions for each specific condition.

Table 1 shows the regular expressions used to detect tweets related to the analysed diseases in Spanish. A similar list was used for Portuguese.

The use of regular expressions allowed obtaining large sets of tweets related to the specified diseases. However, among the obtained tweets, negative sentences that do not indicate the presence or absence of a disease in one person, such as "Hoping the flu does strike me again this winter", may also occur . To solve this problem, we applied machine learning techniques on the datasets obtained using the regular expressions, in order to filter such cases (see Section 3.1). This allowed differentiating the tweets that

Table 1. Regular expressions for detecting health disorders in Spanish tweets

Flu	Regular Expression
Flu	$(grip[a-z]+)$
Cold	$(costip[a-z]+)$
	$(fiebre.*grado(s)?)\|(grado(s)?.*fiebre)$
Flu Symptoms	$(((dolor(es)?\|(meduele)).*(cuerpo\|cabeza\|garganta).*fiebre)\|$
	$(fiebre.*((dolor(es)?\|(meduele)).*(cuerpo\|cabeza\|garganta)))$

Pregnancy	
Pregnancy	$(embaraz[a-z]+)$
Common phrases	$(espero\|tendre).*((un\|una\|unos\|unas)?s(hij[a-z]+\|niño(s)?\|bebe(s)?\|niñit[a-z]+))$
	$(ser(e)?\|soy\|somos).*(padre(s)?\|madre)$

Depression	
Depression	$(depres[a-z]+)$
Depressed	$(deprim[a-z]+)$
Common phrases	$((problema(s)?\|disturbio(s)?).*(mental\|mentales\|psicologico(s)?\|psiquiatrico(s)?))$
	$(quiero).*(morir\|morir[a-z]+)$
	$(todo(s)?).*(dia(s)?).*(trist[a-z]+\|problema(s)?)$

Eating Disorders	
Obesity	$(obesidad\|obeso\|obesa)$
Overweight	$sobrepeso$
Bulimia	$(bulimia\|bulimica\|bulimico)$
Anorexia	$(anorexia\|anorexica\|anorexico)$
Bigorexia	$(vigorexia\|vigorexica\|vigorexica)$
Ebigorexia	$(ebigorexia\|ebigorexica\|ebigorexico)$
Orthorexia	$(ortorexia\|ortorexica\|ortorexico)$
	$(hacer\|hago\|hice).*((dieta(s)?\|regime(s)?)\|(dieta(s)?.*regime(s)?))$
	$(soy\|estoy\|mesiento).*gordo$
Common phrases	$((no)\|(quier[a-z]+)).*engordar$
	$(excesso\|problema\|riesgo\|peligro).*peso$
	$(enfermedad(es)?\|problema(s)?).*aliment[a-z]+$

Table 2. Number of features for each dataset

	Flu	Depression	Pregnancy	Eating Disorders
Spanish Tweets	608	721	698	567
Portuguese Tweets	842	983	1042	747

only mention a given disease from those which actually indicate that the person has the disease.

3.1 Machine Learning

To apply machine learning we need to obtain a set of features from the subsets of tweets related to the studied diseases (obtained by applying regular expressions (see Table 3). For that, we represented these tweets in a bag-of-words (BOW) model after removal of stopwords[3] and word stemming [13]. Character bigrams were also included in the feature set. According to the language and the disease studied, we obtained different sets of features, as shown on Table 2.

In order to identify the best classifier to our method, we have tested the obtained features with various machine learning techniques (SVM, Naïve Bayes, Decision Trees and Nearest Neighbour). To test these techniques, we used WEKA [8], an open source tool for data mining and machine learning that includes multiple implementations of different existing techniques.

---

[3] http://snowball.tartarus.org/

Table 3. Datasets used in the experiments

	Flu	Depression	Pregnancy	Eating Disorders
Spanish Tweets	827	3253	1985	412
Portuguese Tweets	1150	2845	2626	455

4   Experimental Results

In this section, we explain the datasets used in the experiments and several issues about how the experiments were made. We then analyse and discuss the results obtained with the proposed method for each disease.

4.1   Experimental Setup

To acquire the tweets for this study, we developed an application that uses the Twitter search API [18] and the geocoding information contained in the tweet metadata to obtain only tweets originated in Spain and Portugal. Furthermore, in order to filter out tweets not written in Spanish or Portuguese, we used the "language detector" library [16]. This library is based on Bayesian filters and has a precision of 0.99 in detecting the 53 languages it supports. Tweets were acquired during 30 days (from October 30th to November 30th, 2012).

The Spanish and Portuguese datasets contain approximately 5.8 and 4.5 million tweets, respectively. Table 3 shows the number of tweets considered for each language and disease pair, after applying the regular expressions shown in Table 1. The filtered tweets were manually labelled to be used for testing the machine learning algorithms. A tweet is considered true when it indicates the presence of one of the studied diseases in the person who have wrote the tweet. In any other case the tweet is considered false.

For the evaluation of the classifier we used a ten-fold cross validation technique. We used a polynomial kernel with $C = 1.0$, for SVM, and the default WEKA parameters for the remaining methods.

4.2   Results

Using all the features of each disease calculated for each country, we tested different implementations of the classifiers. The results for each type of classifier are shown on Table 4.

In the results shown, the Naïve Bayes classifier achieved the best results in all the cases except for 'Flu' in Spanish tweets. This classifier obtained in many cases a precision and a recall higher than 0.9, with an AUC always higher than 0.7, and often near 0.9. The second best classifier was the Decision Tree, followed by kNN. The worst results were obtained with the SVM classifier, with an AUC below 0.7 in some cases.

On the other hand, we can see that the best results were obtained in depression and in pregnancy (in Portuguese tweets). Regarding the country, in general, better results were obtained in the Portuguese dataset.

Table 4. Results obtained on the datasets. AUC = Area Under the receiver operating characteristic Curve.

Disease	Classifier	Spanish Tweets				Portuguese Tweets			
		F-Measure	Precision	Recall	AUC	F-Measure	Precision	Recall	AUC
Depression	Naïve Bayes	0.913	0.949	0.891	**0.878**	0.912	0.947	0.887	**0.833**
	SVM	0.946	0.948	0.944	0.739	0.902	0.934	0.876	0.691
	Decision Tree	0.976	0.968	0.985	0.845	0.974	0.963	0.985	0.762
	kNN	0.862	0.937	0.814	0.784	0.900	0.937	0.871	0.768
Pregnancy	Naïve Bayes	0.952	0.948	0.957	**0.703**	0.977	0.973	0.982	**0.877**
	SVM	0.940	0.942	0.939	0.644	0.945	0.975	0.920	0.679
	Decision Tree	0.947	0.944	0.951	0.689	0.978	0.971	0.985	0.801
	kNN	0.949	0.945	0.953	0.701	0.979	0.975	0.985	0.714
Flu	Naïve Bayes	0.766	0.759	0.775	0.743	0.667	0.667	0.669	**0.746**
	SVM	0.755	0.749	0.764	0.696	0.681	0.691	0.690	0.671
	Decision Tree	0.749	0.757	0.804	0.670	0.672	0.672	0.674	**0.746**
	kNN	0.761	0.756	0.799	**0.786**	0.687	0.687	0.689	0.745
Eating Disorders	Naïve Bayes	0.720	0.720	0.720	**0.714**	0.786	0.785	0.817	**0.744**
	SVM	0.683	0.688	0.679	0.607	0.725	0.729	0.720	0.650
	Decision Tree	0.785	0.756	0.817	0.630	0.869	0.838	0.902	0.690
	kNN	0.684	0.714	0.669	0.696	0.667	0.737	0.630	0.686

5 Conclusions

This article presents a method to extract a set of tweets that show the presence of certain diseases (flu, depression, pregnancy, eating disorder) in the society. The study was centred in Spain and Portugal, based on the geocoded data and on the language of the tweets. Using these sets of tweets we aim to measure the presence and evolution of a certain disease in society.

The proposed method is divided into two stages. First, we continuously gathered all tweets of each country and then filtered these tweets by means of several regular expressions, defined specifically for each disease. Secondly, we used machine learning methods, specifically Naïve Bayes, SVM, Decision Trees and kNN classifiers, in order to remove false positive documents identified with the regular expressions.

Compared to previous works, the main advantages proposed in this study are the detection of several health conditions in two distinct languages. The results obtained are very promising and indicate that such an approach provides a feasible solution for measuring and tracking the evolution of many health related aspects within the society.

Finally, we want to highlight the results obtained by our method applying Naive Bayes, which has obtained a precision and a recall close to 0.9. Based on this fact, we present Naive Bayes as the most suitable classifier for the proposed method to detect diseases in Twitter.

6 Future Work

Other types of user-generated content, such as Internet searches or comments to news articles, may also contain information related to some of these aspects. Thus, this information could be used to complement the data extracted from Twitter.

The proposed method may be extended to other languages and subjects, providing a continuous monitoring system of health pandemic or social issues, in a larger geographic region.

Acknowledgements. This research was supported by Xunta de Galicia CN2012/211, the Ministry of Education and Science of Spain and FEDER funds of the European Union (Project TIN2009-14203) and by "Fundação para a Ciência e a Tecnologia" (FCT, Portugal) under project PTDC/EIA-CCO/100541/2008 and Ciência2007 programme.

References

1. Aramaki, E., Maskawa, S., Morita, M.: Twitter catches the flu: detecting influenza epidemics using Twitter. In: Proceedings of the Conference on Empirical Methods in Natural Language Processing, pp. 1568–1576. Association for Computational Linguistics (2011)
2. Bosley, J.C., Zhao, N.W., Hill, S., Shofer, F.S., Asch, D.A., Becker, L.B., Merchant, R.M.: Decoding twitter: Surveillance and trends for cardiac arrest and resuscitation communication (2012)
3. Chew, C., Eysenbach, G.: Pandemics in the age of Twitter: content analysis of Tweets during the 2009 H1N1 outbreak. PloS one 5(11), e14118 (2010)
4. Chunara, R., Andrews, J.R., Brownstein, J.S.: Social and News Media Enable Estimation of Epidemiological Patterns Early in the 2010 Haitian Cholera Outbreak. American Journal of Tropical Medicine and Hygiene 86(1), 39–45 (2012)
5. Culotta, A.: Towards detecting influenza epidemics by analyzing Twitter messages. In: Proceedings of the First Workshop on Social Media Analytics, pp. 115–122. ACM (2010)
6. Culotta, A.: Detecting influenza outbreaks by analyzing Twitter messages, arXiv:1007.4748 [cs.IR] (2010)
7. Ginsberg, J., Mohebbi, M.H., Patel, R.S., Brammer, L., Smolinski, M.S., Brilliant, L.: Detecting influenza epidemics using search engine query data. Nature 457, 1012–1014 (2009)
8. Hall, M., Frank, E., Holmes, G., Pfahringer, B., Reutemann, P., Witten, I.H.: The weka data mining software: an update. SIGKDD Explor. Newsl. 11, 10–18 (2009)
9. Heaivilin, N., Gerbert, B., Page, J.E., Gibbs, J.L.: Public health surveillance of dental pain via Twitter. Journal of Dental Research 90(9), 1047–1051 (2011)
10. Lampos, V., Cristianini, N.: Tracking the flu pandemic by monitoring the social web. In: 2010 2nd International Workshop on Cognitive Information Processing (CIP), pp. 411–416 (2010)
11. Lyon, A., Nunn, M., Grossel, G., Burgman, M.: Comparison of web-based biosecurity intelligence systems: BioCaster, EpiSPIDER and HealthMap. Transboundary and Emerging Diseases 59(3), 223–232 (2012)
12. Paul, M., Dredze, M.: You are what you tweet: Analyzing Twitter for public health. In: Proceedings of the 5th International AAAI Conference on Weblogs and Social Media, pp. 265–272 (2011)

13. Porter, M.F.: Snowball: A language for stemming algorithms. (published online, October 2001)
14. Santos, J.C., Matos, S.: Predicting Flu Incidence from Portuguese Tweets. In: Proceedings of IWBBIO 2013, Granada, Spain (March 2013)
15. Scanfeld, D., Scanfeld, V., Larson, E.L.: Dissemination of health information through social networks: twitter and antibiotics. American Journal of Infection Control 38(3), 182–188 (2010)
16. Shuyo, N.: Language detection library for java (2012)
17. Signorini, A., Segre, A.M., Polgreen, P.M.: The use of Twitter to track levels of disease activity and public concern in the U.S. during the influenza A H1N1 pandemic. PloS One 6(5), e19467 (2011)
18. Twitter search api (2012), https://dev.twitter.com/docs/api/1/get/search (online; accessed November 20, 2012)

Optimized Workflow for the Healthcare Logistic: Case of the Pediatric Emergency Department

Inès Ajmi[1], Hayfa Zgaya[2], and Slim Hammadi[1]

[1] LAGIS/EC-Lille UMR CNRS 8219
[2] EA 2694, Laboratory of Public Health

Abstract. The Emergency Department (ED) in a hospital, as its name implies, is a facility to be utilized by those who require emergency medical care. This paper introduces the longitudinal organization of the patient handling" in the Pediatric Emergency called the "Pediatric Emergency Path". This work discusses the usability of the workflow approach in order to design the patient path in the Pediatric Emergency Department (PED) in order to thwart the care complexity scheme. The goal is to optimize these paths to improve the quality of the patient handling while mastering the wait time. The development of this model was based on accurate visits made in the PED of the Regional University Hospital Center (CHRU) of Lille (France). This modeling, which has to represent most faithfully possible the reality of the PED of CHRU of Lille, is necessary. It must be enough retailed to produce an analysis allowing to identify the dysfunctions of the PED and also to propose and to estimate prevention indicators of tensions. Our survey is integrated into the French National Research Agency project, titled: "Hospital: optimization, simulation and avoidance of strain" (ANR HOST)[1].

Keywords: PED, workflow, modeling, peak of activity, tension indicators.

1 Introduction

In France, as in many countries of Europe, the emergency departments present the same difficulties whose reasons are multiple. These difficulties are not only linked to the health organization but also to the evolution of the western civilization [1]. The requirements concerning health care evolved in relation with a new approach to management of the time. Today, people require a fast and efficient handling. They reject the ageing, the illness and the death [2]. The arrival patient flow to the emergency department keeps increasing. This rise has generated a strategic interest in optimizing the technical and human resources while mastering the costs [3]. In order to reach these objectives, the health establishments have resorted to the tools and techniques of management borrowed from the industry domain [4] as the Workflow tool that will be used in this article. The use of Workflow methodology showed applicability and the interest of the company modeling method to reorganize a health

---

[1] ANR-11-TecSan-010.

establishment. It allowed improving the performance of different service and activities conduct system [4]. The Discrete Event Simulation (DES) techniques have been used a lot for modeling the operations of an Emergency Department (ED). The model was developed to test alternative ED attending physician-staffing schedules and to analyze the corresponding impacts on patient throughput and resource utilization [5], to help the ED managers understand the behavior of the system with regard to the hidden causes of excessive waiting times [6], to analyze patient flows and throughput time [7] [8][9][10]. DES has also been used for estimating future capacities of new ED facilities or expansions [11] [12]. The main objectives of the present paper are: 1) To model the Pediatric Emergency Department (PED) using Workflow Methodology for better understanding the patient flow process through the PED, 2) To simulate using the same tool Workflow in order to identify and analysis the dysfunctions of the PED and also to propose and to estimate prevention indicators of tensions.

This article is structured in five sections. The first one presents the context of the survey. We introduce the notion of "Emergency Path" in section two. In the third section, we explain the workflow modeling at the PED Section four describes the simulation model and some experiments and results. Finally in section five we draw the conclusions of this work.

2   Context of Survey

The health system is characterized by an increasing complexity which requires the performance and the mastery of the costs, and so comes the necessity to have new tools in support of the management strategy. This problematic has generated the idea of the HOST ANR national project in the PED which has the objective to elaborate a new methodological approach for the anticipation of the tensions of the complex care production system and more especially of the emergency paths. The problematic of strategy, the performance assessment and traceability of the patient course are integrated into the approaches which are suggested. The HOST ANR national project is a scientific research work in the hospital world with fallouts for the PED of the CHRU of Lille.

The scientific fallouts expected of this project are:

- A prospective vision of the conception and the piloting of global emergency activity handling system in European context.
- A good tool to anticipate the tensions of the PED.
- The fallouts for emergency paths actors will consist in establishing the methods and operational tools allowing:
- To bring   the recommendations for the emergency paths conception and reengineering for health establishments,
- To improve the handling efficiency and the quality of the service returned to the patient,

- To assure a better internal and external coordination with the other actors of the emergency path when the tensions can't be avoided.
- To permit best practices sharing.

Our work represents the first step of the national ANR HOST project realization and discusses the emergency handling system. The ubiquity of the emergency problem in France and in Europe, the strategic place of the ED in the health establishments, the frequency and the huge media coverage of the dissatisfaction of the ED users and rising health care costs are primarily responsible for the high level of interest in PED. The PED of the CHRU of Lille nursing staff constituted by physicians, surgeons, senior executives, nurses…etc. and administrative staffs of the CHRU of Lille are implied in our survey. The direction of the quality, strategy, and the medical information participate in the realization of this work.

3   The Emergency Path in the PED

Emergency Path is a longitudinal organization of the patient's handling. It is not a structure but an operative concept. The patient flow can vary from patient to patient based on acuity level and diagnosis [12]. The hospital establishments are chronically confronted to a problematic for which no satisfying answer exists at the present time. This problematic is resulting from the permanent interferences between the programmed and the non-programmed activity, and more especially the urgent non programmed activity. We can define two operational concepts that will intervene in the description of the PED: 1) The incoming flows: programmed or non-programmed, with almost periodic or uncertain variation, 2) The retiring flows: constituting the downstream of the PED. The emergency path, the incoming and the retiring flows are three concepts that describe the PED like a complex system in interaction. We proceeded to the modeling of the survey project at the PED of the CHRU of Lille. The phases of Workflow of the PED modeling are:

- Description of the features of each element of the global process and its subprocesses in the PED,
- Modeling the PED (flow and resources organization) and its interactions with the other internal components of the CHRU of Lille,
- To define a typology of the patients admitted in the PED of the CHRU of Lille.

4   The PED Workflow Modeling

The main objective of our survey is to identify the dysfunctions of the PED and also to propose and to estimate prevention indicators of tensions. To reach this objective, it is necessary to have a global Workflow model of the PED and its environment. Before optimizing and simulation of the PED, it is necessary to analyze and to characterize the PED structure. It requires the use of Workflow methodology to represent the functional and process view of the PED and its related parts.

4.1   The PED of the CHRU of Lille

The term of emergency covers two distinct phenomena:

- The recurrent flows are able to present some seasonal variations, but known on average on middle horizons (i.e. month or year). Even, if these flows are feared, quantified in volume and in nature on medium and short term, the setting up of a structure, an organization and a piloting is a major stake of efficiency of the care production.
- The flows following upon the sanitary crises (flow, heat waves, cold weather waves). In this case, the flows are completely unforeseen in volume and in nature.

The adaptation and the implementation of organization, scheduling, management and optimization approved methods are foreseeable for the programmed patients flows handling , but it is a lot more complex to master the emergency flows handlings. The problem is not simple because today the emergencies are considered by the actors of the health system themselves as main entropy generator of the cares system.

4.2   The Workflow Methodology

To model the health system, a lot of company modeling methods has been used. The company modeling is the representation of the enterprise in terms of strategy, structure, functionalities, behavior, organization, evolution and relations with the environment. One of these methods is the Workflow methodology. It has been used for modeling, diagnosis and conception of a hospital system [13]. This work permitted to demonstrate applicability and the interest of the company modeling methods to reorganize a health establishment. The objective of the development and description of the workflow model is to assess the overall processing capabilities of the flow in order to support various joint activities between medical staff that is temporally and spatially dispersed. Control of the workflow for the purpose of optimizing the placement of limited medical resources, both personnel and equipment, on the medical scene is an important issue. Since typical unforeseen circumstances will frequently occur in the PED environment, there is a need for an effective model, capable of dynamic control in workflow descriptions, for medical treatment. Thus, for descriptions of the workflow in this paper, we have used the BPMN standard graphical language, which is easy for users to express and is readily comprehensible [14].

4.3   The Phase of Modeling

We are going to model all paths exist at PED of the CHRU Lille. We are going to represent the physical system service and the PED to have a complete idea of the patient flows. In this paper, we start with the representation of the physical system in the PED and the emergency paths of External care, Unit of short term hospitalization and Care emergency vital. We represent only the models that will be analyzed after modeling the whole paths of the PED of the CHRU Lille in the ulterior stages of our survey. The physical systems of the PED and the emergency paths that we have just mentioned previously are represented by the figures1, 2, 3 and 4. The set of the functional models presented in this paper have been achieved with the Bonita soft software Workflow. We remark that these models are characterized by the diversity of the activities and the big number of people that intervenes in the patient handling process and its diagnosis nature. We note in more the uncertain criteria in the patients handling and the risks that can emerge during his path. The aggravation of a patient's state can already stimulate the change of his cares process anticipated. It confirms us the complexity of the emergency handling system. It is also interesting to see that some activities of these processes are decisional. These decisions are not management decisions but the choices in the patient [15] (Care traditional hospitalization, external care, Unit of short term hospitalization, Care as a matter of emergency vital, Consultation in a cubicle).

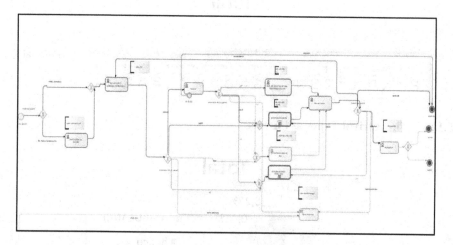

5   Emergency Department Simulations

After modeling the emergency department, we intend to use discrete event simulation techniques to conceive a simulation model which is comprehensible, flexible, and easy for use and representative of the reality. Since, we consider using this model in the ulterior stages of our solution and to adapt it to the different patient emergency paths (Care traditional hospitalization, external care, Unit of short term hospitalization, Care as a matter of emergency vital, Consultation in a cubicle).

5.1   The Data Analysis

To do the simulations, we need the patients' information. We could collect some information at the PED of CHRU Lille which allowed studying the different features of the patient.

5.1.1   The Sex

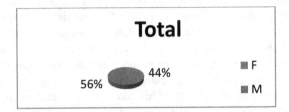

5.1.2   The Arrival Flow of Patient at the PDE

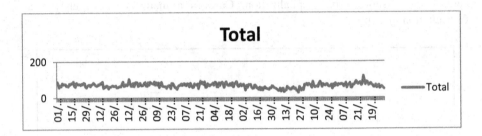

5.1.3   The Arrival Types at the PED

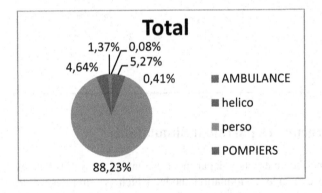

5.1.4   The Age

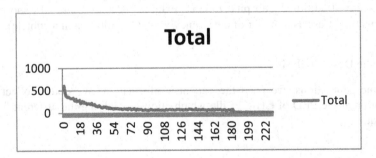

5.2   Experiments and Results

For a simulation on last year, the main gotten results are presented in Table 1 simulation results and Table 2 global statistics. It is possible to look at other numerous statistics but it would constitute a deeper analysis that will be treated in the ulterior stages of our survey.

Table 1. Simulation results

Simulation	NOR	Adm. Recept	Amb	No Amb	ZVSLH	Isol.Room	VERR
Operative Times (minutes)	7	10	30	30	420	300	50
Sizes averages of the waiting lines (in the passageway and in	0	3,0366	0,1153	0	6,3022	0,0058	0
Middle rate occupancy of resources (in %)	53,958	76,994	33,326	25,771	94,692	19,989	8,507

Table 2. Global statistics

Global statistics	
NbEntry	3330
NbOutlays	3312
Arrived Cadence	13.032
Middle holding time	331.92

6   Conclusion and Prospects

In this paper, we presented the setting of our survey that will last two years in the PED of CHRU Lille, the first step of the modeling phase of the patient handling system in the PED and the PED simulation .This first step allowed us to conclude that the simulation model developed in this paper can be used to help identify process inefficiencies and to evaluate the effects of layout, staffing, resource and patient flow changes on system performance. In the ulterior stages of this survey, we are going to model the totality of the emergency paths in the PED of CHRU of Lille, to analyze the gotten models and to conceive a complete model of the patient handling system. We are going to achieve a simulation of these models to value the modeling system performances (percentage occupancy of resources, service quality indicators).

References

1. Reix, A.: Etude de l'organisation des différents systèmes d'urgence dans six pays européens: La France, le Royaume Uni, la Belgique, la Suède, la Finlande et la Norvège. Memory of thesis for the diplôma of doctor state in medicine: 18 (2002)
2. Chodosas, M.: Etude de l'organisation des différents systèmes d'urgence dans cinq pays européens: La France, l'Allemagne, l'Espagne, l'Italie et le Portugal. Mémoire de thèse pour le doctorat en médecine: 20 (2002)
3. Navas, J.-F., Artéta, C., Hadjes, P.S., Jiménez, F.: « Construction et simulation d'un modèle de fluxde patients dans le service d'urgences d'un hôpital colombien «, Formato de Atencion Inicial de Urgencias, en « Procesos Operativos de Urgencias », Documentos confidenciales Servicio de Urgencias Fundacion Cardiolnfantil, Agosto (2003)
4. Ducq, Y., Vallespir, B., Doumeingts, G.: Utilisation de la méthodologie GRAI pour la modélisation, le diagnostic et la conception d'un système hospitalier. GISEH

5. Rossetti, M.D., Trzcinski, G.F., Syverud, S.A.: Emergency department simulation and determination of optimal attending physician staffing schedules. In: Farrington, P.A., Nembhard, H.B., Sturrock, D.T., Evands, G.W. (eds.) Proceeding of the 1999 Winter Simulation Conference (1999)
6. Ruby, B.E., Darrell, S.W., Wendy, A.S., Mary, H.C.: The use of simulation to evaluate hospital operations between the emergency department and a medical telemetry unit. In: Chick, S., Sanchez, P.J., Ferrin, D., Morrice, D.J. (eds.) Proceedings of the 2003 Winter Simulation Conference, pp. 1887–1893 (2003)
7. Simon, S., Wendy, A.S., Darrell, S.W.: The use of simulation to reduce the length of stay in an emergency department. In: Chick, S., Sanchez, P.J., Ferrin, D., Morrice, D.J. (eds.) Proceedings of the 2003 Winter Simulation Conference, pp. 1907–1911 (2003)
8. Mahapatra, S., Koelling, C.P., Patvivatsiri, L., Fraticelli, B., Eitel, D., Grove, L.: Pairing emergency severity index5-level triage data with computer aided system design to improve emergency department access and throughput. In: Chick, S., Sanchez, P.J., Ferrin, D., Morrice, D.J. (eds.) Proceedings of the 2003 Winter Simulation Conference, pp. 1917–1925 (2003)
9. Soemon, T., Hiroko, S.: Functional analysis for operating emergency department of a general hospital. In: Ingalls, R.G., Rossetti, M.D., Smith, J.S., Peters, B.A. (eds.) Proceedings of the 2004 Winter Simulation Conference, pp. 2003–2011 (2004)
10. Felipe, B.F., Hector, J.E., Mahal, D.C.: The use of simulation and design of experiments for estimating maximum capacity in an emergency room. In: Chick, S., Sanchez, P.J., Ferrin, D., Morrice, D.J. (eds.) Proceedings of the 2003 Winter Simulation Conference, pp. 1903–1906 (2003)
11. Allan, W., Rainer, D.: Using simulation in the architectural concept phase of an emergency department design. In: Chick, S., Sanchez, P.J., Ferrin, D., Morrice, D.J. (eds.) Proceedings of the 2003 Winter Simulation Conference, pp. 1912–1916 (2003)
12. Rossetti, M.D., Trzcinski, G.F., Syverud, S.A.: Emergency department simulation and determination of optimal attending physician staffing schedules. In: Farrington, P.A., Nembhard, H.B., Sturrock, D.T., Evands, G.W. (eds.) Proceeding of the 1999 Winetr Simulation Conference (1999)
13. Souf, N.-B., Renard, J.-M., Régis: Dynamic workflow model for complex activity in intensive care unit. International Journal of Medical Informatics 53, 143–150 (1999)
14. Komashie, A., Mousavi, A.: Modeling emergency departments using Discrete Event Simulation techniques. In: Kuhl, M.E., Steiger, N.M., Armstrong, F.B., Joines, J.A. (eds.) Proceeding of the 2005 Winter Simulation Conference (2005)
15. Fraisse, A., Le Bel, S., Mas, B., Macrae, D.: Paediatric cardiac intensive care unit: Current setting and organization. Science Direct, 546–551 (2010)

Boosting the Detection of Transposable Elements Using Machine Learning

Tiago Loureiro[1], Rui Camacho[2], Jorge Vieira[3], and Nuno A. Fonseca[4]

[1] DEI & Faculdade de Engenharia, Universidade do Porto, Portugal
tiagodloureiro@gmail.com
[2] DEI & Faculdade de Engenharia & LIAAD-INESCTEC, Universidade do Porto, Portugal
rcamacho@fe.up.pt
[3] IBMC - Instituto de Biologia Molecular e Celular & Universidade do Porto, Portugal
jbvieira@ibmc.up.pt
[4] EMBL Outstation, European Bioinformatics Institute (EBI), Hinxton,
Cambridge CB10 ISD, UK
CRACS-INESCTEC, Portugal
nf@ebi.ac.uk

Abstract. Transposable Elements (TE) are sequences of DNA that move and transpose within a genome. TEs, as mutation agents, are quite important for their role in both genome alteration diseases and on species evolution. Several tools have been developed to discover and annotate TEs but no single one achieves good results on all different types of TEs. In this paper we evaluate the performance of several TEs detection and annotation tools and investigate if Machine Learning techniques can be used to improve their overall detection accuracy. The results of an *in silico* evaluation of TEs detection and annotation tools indicate that their performance can be improved by using machine learning classifiers.

Keywords: Transposable Elements, Machine Learning, Genomics.

1 Introduction

Transposable Elements (TE), also known as transposons, are sequences of DNA that move and transpose within a genome. TE's role as mutation agents is important in both genome alteration diseases and on species evolution [3] [4] [12] [8][9][2]. Several methods have been developed to discover and annotate Transposable Elements. In [1] an extensive list of TE detection methods is surveyed. These methods have been classified in four main categories [1]: *De novo*; Structure-based; Comparative Genomic; and Homology-based. Although there are different tools, based on these methodologies, for detecting transposable elements there is not any single tool achieving good results on different types of TEs.

In this paper we evaluate existing TE detection tools using *in silico* data and study if Machine Learning techniques can be used to combine several TE detection tools predictions in order to improve the overall detection accuracy.

The remainder of the paper is organised as follows. Section 2 explains the data generation process for evaluation the TEs detection tools. In Section 3 we present an empirical evaluation of the TEs detection tools. In Section 4 we explain the ML experiments

M.S. Mohamad et al. (Eds.): *7th International Conference on PACBB*, AISC 222, pp. 85–91.
DOI: 10.1007/978-3-319-00578-2_12   © Springer International Publishing Switzerland 2013

to improve the performance of detecting TEs. Finally, in Section 5, we draw some conclusions.

2   *In Silico* Data

In order to evaluate the TE detection tools it is essential to have curated data sets of genome sequences. Table 1 summarizes the data used (namely TEs, genes, repetitive elements, etc) to assemble 'artificial' sequences. The set of 'real' genes was obtained from FlyBase [13] (Drosophila melanogaster). The 'real' TEs were obtained from Repbase [7] and from Gydb [11]. Other variables considered in the simulation were mutations, either point mutations or *indel* mutations, the length, composition and abundance of TE.

Table 1. Data used to produce simulated sequences

	Element type	Number
	Autonomous LTR Retrotransposons	2248
	Non autonomous LTR Retrotransposons	379
	DIRS	14
	Non-LTR Retrotransposons	140
	Autonomous non-LTR Retrotransposons	604
TEs	Non autonomous non-LTR Retrotransposons	384
	TIR	1247
	DNA Transposons	628
	Helitrons	139
	Politrons	24
	Genes	15458
	Repetitive Elements	147

The simulation parameters included the length of the sequence to be produce, percentage of genes included in the sequence in relation to its total length, percentage of TEs included in the sequence in relation to its total length, percentage of repetitive elements (no transposons included) that should be included in the sequence in relation to its total length. rate of insertions, deletions and replacements. Producing sequences using different combinations of these parameter's values allowed us to generate a diverse set of DNA sequences. The output data of a simulation is a set of sequences, written in FASTA [10] format, and an annotation file containing all TEs and repetitive elements locations inside each sequence. A simulated sequence consists of genes, transposons and other repetitive elements filled with random nucleotides in the gaps between them. The quantity of TEs, genes and repetitive elements are defined by the parameters referred above.

3   Evaluation of Transposon Detection Tools

Each TE detection tool analyzes all the sequences (of a given data set) and produces as a result the annotations of the TEs. The general accuracy was computed based on the predicted location of TEs and the "true" locations generated by the simulator.

In this study we have evaluated five tools that we next briefly describe.

PILER[1] [5] is a *de novo* TE detection tool that adopts a heuristic-based approach to *de novo* repeat annotation that exploits characteristic patterns of local alignments induced by certain classes of repeats. The PILER algorithm is designed to analyze assembled genomic regions and find only repeat families whose structure is characteristic of known subclasses of repetitive sequences.

BLAT [15] is a mRNA/DNA alignment tool. It uses an index of all non-overlapping K-mers in a given genome to find regions likely to be homologous to the query sequence. It performs an alignment between homologous regions and stitches together these aligned regions into larger alignments.

CENSOR[2] [6] was designed to identify and eliminate fragments of DNA sequences homologous to any chosen reference sequences. It uses BLAST to identify matches between input sequences and a reference library of known repetitive sequences. The length and number of gaps in both the query and library sequences are considered along with the length of the alignment in generating similarity scores. This tool reports the positions of the matching regions of the query sequence along with their classification.

RepeatMasker[3] [14] discovers repeats and removes them to prevent complications in downstream analysis sequence assembly and gene characterization. Identification of repeats by RepeatMasker is based entirely upon shared similarity between library repeat sequences and query sequences. The output of the program is a detailed annotation of the repeats that are present in the query sequence as well as a modified version of the query sequence in which all the annotated repeats have been masked.

LTR_Finder[4] [17] predicts locations and structure of full-length LTR retrotransposons accurately by considering common structural features. LTR_FINDER identifies full-length LTR element models in genomic sequence. This program reports possible LTR retrotransposons models at different confidence levels.

In Table 2 the average accuracy of each tool regarding each TE type is presented. LTR_Finder achieved poorer results in finding most types of TEs. Overall both Censor and RepeatMasker were the most accurate tools in finding different types of TEs.

4   Machine Learning to Improve TEs Detection Tools

Based on the experimental results of the TEs detection tools evaluation (Section 3) we have investigated if Machine Learning (ML) algorithms could improve TEs detection. We have used a two step process for TEs detection using ML: i) determine if a certain item (subsequence) in the sequence is or not a TE (TE detection); and ii) if the item has been classified as a TE then we determine its boundaries (TE annotation). The first step is concerned with the choice of the best tools to identify a TE with some given characteristics. The second step aims at choosing a tool that minimizes the error of an inferred TE boundary.

---

[1] http://www.drive5.com/piler/

[2] http://www.girinst.org/downloads/software/censor/

[3] http://www.repeatmasker.org/

[4] http://tlife.fudan.edu.cn/ltr_finder/

Table 2. Accuracy (%) per TE type

Tool	Aut. LTR Retrotransposons	Non Aut. LTR Retrotransposons	DIRS	Non-LTR Retrotransposons	Aut. Non-LTR Retrotransposons	Non Aut. Non-LTR Retrotransposons	TIRS	DNA TEs	Helitrons	Politrons	All
BLAT	26.69	18.12	2.72	23.12	19.68	37.69	20.46	13.67	21.92	11.14	19.61
Censor	**61.49**	**82.68**	**81.43**	**71.1**	**74.02**	**68.45**	**78.85**	**82.9**	**52.13**	20.86	**67.38**
LTR_Finder	0.17	0.22	0.1	0.02	0	0	0	0	0	0	0.05
PILER	0.51	38.05	36.33	37.46	46.6	10.24	28.27	25.63	41.94	**23.56**	28.66
Repeat Masker	51.66	58.1	31.1	43.88	51.71	55.63	57.66	57.14	42.23	4.62	45.28

The *Rapidminer* [5] software which uses *Weka* [16] algorithm implementations was used to build the classifiers. The algorithms considered: *Weka*'s implementation of Neural networks using 500 training cycles and 0.3 of learning rate; Bayes Network; Random Forest classifier to build an ensemble of decision trees; Decision Trees based on the C4.5 algorithm. The classifiers performance was estimated by measuring the accuracy in a 10 fold cross-validation procedure.

The classification of a potential TE candidate as a TE or not is a typical classification problem. In these terms, we used a data set containing 325000 examples, equally distributed in terms of TE types and in terms of being real TEs or false positives. The features used as the input for the models were the discretized TE length (using Equal-depth Binning in 50 categories), the TE type, the tool that made the prediction (FOUNDTOOL), and a IS_TE feature as the class. The IS_TE feature is a boolean which indicates whether a given example is or is not a TE. Table 3 shows the results obtained for the different ML algorithms considered. The best results were achieved with Decision Trees with an average accuracy of 98%, although the difference to Random Forest is not significantly different.

Table 3. TE detection: accuracy using different classification algorithms

Algorithm	Accuracy (%)
Neural Network	69.01
Naive Bayes Net	96.30
Random Forest	98.90
Decision Trees	98.92

---

[5] http://www.rapidminer.com/

The sensitivity of the tools for the level of mutations present in the sequences analyzed was also a theme that we wanted to clarify. The results (not shown) suggest that BLAT and PILER tools are influenced by mutations present in the DNA sequences. On the other hand, the performance of Censor, LTR_Finder and RepeatMasker were not affected significantly by the level of mutations.

Finding the Best TE Annotation Tool. Which tool minimizes the predicted location error for a given TE candidate? To answer this question we used a set of 129 198 examples of TE elements equally distributed between the different TE classes. "bestTool" is the class label and we have used the following features: TE type; set of tools that have detected the TE in step1; number of such tools that have detected the TE in step 1; and the class of the tools that have detected the TE in step 1. The bestTool feature is the name of the tool with the minimum location error.

We tested different model generation algorithms, all subjected to a 10 fold cross-validation process, to assess their performance. In Table 4 the results obtained with the different learner algorithms are compared. Again, the model with highest accuracy was produced with Decision Trees. Table 5 presents the confusion matrix of this model. This classifier has a high accuracy and can perform well with the tested artificial data. It is also worth to mention that the LTR_Finder tool was never used in this context as the location error performance of this tool is considerable lower than the others.

Applying machine learning to construct classifiers in the TE detection scope can further improve the accuracy of TE detection and annotation. In all the different problems, the approach that produced best results was Decision Trees (W-J48 Weka implementation).

Table 4. TE annotation: Classification algorithms model comparison. ZeroR measures the majority class percentage and is used as a base line value.

Algorithm	Accuracy (%)
Ridor	96.43 (0.10)
Naive Bayes Net	96.37 (0.18)
Random Forest	96.56 (0.14)
Decision Trees	96.56 (0.14)
ZeroR	76.55

5 Conclusions

In this paper we have assessed a set of computational tools for detecting Transposable Elements. The results obtained suggest that both Censor and RepeatMasker are the most accurate tools in detecting TEs. In a particular category, Politron TEs, the PILER tool obtained the best results. The LTR_Finder tool has achieved, by far, the worse results in this comparison with very low accuracy in the detection of TE. BLAT and RepeatMasker had some problems detecting DIR TEs. On the other hand, Censor scored exceptionally well in this TE category. Politron TEs were also a problem for

Table 5. TE annotation: confusion matrix for the Decision Trees model

	True BLAT	True Censor	True LTR_Finder	True PILER	True RepeatMasker	Class Prediction
Predicted BLAT	98902	0	0	0	0	100.0 %
Predicted Censor	1911	23427	0	0	0	92.5 %
Predicted LTR_Finder	1	0	0	0	4	0.0 %
Predicted PILER	140	71	0	0	85	0.0 %
Predicted RepeatMasker	834	1395	0	0	2428	52.1 %
Class Recall	97.2 %	94.1 %	0.0 %	0.0 %	96.6 %	

tools like RepeatMasker, Censor and BLAT. In this case, PILER performed especially well, outscoring all the other tools.

In terms of inference of TE boundaries, except for the LTR_Finder performance, all the tools performed acceptably well. The biggest issues occurred on the detection of the boundaries of Politron TEs and PILER had some trouble in detecting DIR TEs.

Using different TE tools' predictions from simulated data sets, we generated two classifiers that predict: i) if a given TE candidate is a TE or not, and ii) if it was a TE, predict which tool to use to minimize the boundaries error of that TE.

All in all, we presented evidence that ML models can be used to boost the detection and annotation of existing TE computational tools. Further research is needed to confirm the results in real data.

Acknowledgments. We would like to thank the informatics course "Master in Informatics and Computing Engineering" from Faculdade de Engenharia da Universidade do Porto for providing the conditions for the fulfillment of this work. This work was financed by the ERDF European Regional Development Fund through the COMPETE Programme (operational programme for competitiveness) and by National Funds through the FCT Fundacção para a Ciência e a Tecnologia (Portuguese Foundation for Science and Technology) within projects: FCOMP - 01-0124-FEDER-022701, PTDC/EME-PME/108308/2008 - (COMP- 01-0124-FEDER-010262), and FCOMP-01-0124-FEDER-008916 (PTDC/BIA-BEC/099933/2008).

References

1. Bergman, C.M., Quesneville, H.: Discovering and detecting transposable elements in genome sequences. Briefings in Bioinformatics 8(6), 382–392 (2007)
2. Chénais, B., Caruso, A., Hiard, S., Casse, N.: The impact of transposable elements on eukaryotic genomes: From genome size increase to genetic adaptation to stressful environments. Gene (2012)
3. Casacuberta, E., Gonzlez, J.: The impact of transposable elements in environmental adaptation. Mol. Ecol. (2013)
4. Cowley, M., Oakey, R.J.: Transposable elements re-wire and fine-tune the transcriptome. PLoS Genet. 9(1) (2013)

5. Myers, E.W., Edgar, R.C.: PILER: identification and classification of genomic repeats. Bioinformatics 21, 152–158 (2005)
6. Jurka, J., Klonowski, P., Dagman, V., Pelton, P.: Censora program for identification and elimination of repetitive elements from DNA sequences. Computers & Chemistry 20(1), 119–121 (1996)
7. Jurka, J., Kapitonov, V.V., Pavlicek, A., Klonowski, P., Kohany, O., Walichiewicz, J.: Repbase update, a database of eukaryotic repetitive elements. Cytogentic and Genome Research 110, 462–467 (2005)
8. Kim, Y.J., Lee, J., Han, K.: Transposable elements: No more 'junk dna'. Genomics Inform. 10(4), 226–233 (2012)
9. Koso, H., Takeda, H., Yew, C.C., Ward, J.M., Nariai, N., Ueno, K., Nagasaki, M., Watanabe, S., Rust, A.G., Adams, D.J., Copeland, N.G., Jenkins, N.A.: Transposon mutagenesis identifies genes that transform neural stem cells into glioma-initiating cells. Proceedings of the National Academy of Sciences 109(44), E2998–E3007 (2012)
10. Pearson, W.R., Lipman, D.J.: Rapid and sensitive protein similarity searches. Science 227(4693), 1435–1441 (1985)
11. Llorns, C., Futami, R., Bezemer, D., Moya, A.: The ::::gypsy:::: Database (gydb) of mobile genetic elements. Nucleic Acids Research 36(Database-Issue), 38–46 (2008)
12. Lisch, D.: How important are transposons for plant evolution? Nat. Rev. Genet. 14(1), 49–61 (2013)
13. McQuilton, P., St. Pierre, E., Thurmond, J.: Flybase 101 - the basics of navigating flybase. Nucleic Acids Research 40(Database-Issue), 706–714 (2012)
14. Green, P., Smit, A.F.A., Hubley, R.: RepeatMasker Open-3.0
15. Kent, W.: Blat the blast-like alignment tool. Genome Research 12 (2002)
16. Witten, I.H., Frank, E.: Data Mining: Practical machine learning tools and techniques, 2nd edn. Morgan Kaufmann (2005)
17. Xu, Z., Wang, H.: LTR_FINDER: an efficient tool for the prediction of full-length LTR retrotransposons. Nucleic Acids Research 35(suppl. 2), W265–W268 (2007)

On an Individual-Based Model
for Infectious Disease Outbreaks

Pierpaolo Vittorini and Ferdinando di Orio

Dep. of Life, Health and Environmental Sciences
University of L'Aquila
67100 L'Aquila, Italy
pierpaolo.vittorini@univaq.it

Abstract. The mathematical modelling of infectious diseases is a large research area with a wide literature. In the recent past, most of the scientific contributions focused on compartmental models. However, the increasing computing power is pushing towards the development of individual models that consider the disease transmission and evolution at a very fine-grained level. In the paper, the authors give a short state of the art of compartmental models, summarise one of the most know individual models, and describe a generalization and a simulation algorithm.

Keywords: computational epidemiology, infectious diseases, statistical models, computer simulations.

1 Introduction

Computational epidemiology is a multidisciplinary field that brings together diverse contributions coming from computer science, mathematics, statistics, geographic information science and public health, so to help epidemiologists in their studies concerning e.g. the evolution of epidemics.

In such a context, the mathematical modelling of infectious diseases has a long tradition [10] [12]. Currently, different approaches exist: compartmental models based on differential equations [8] [9], ad-hoc models for the contact process [11][1], or individual-based models [7][4][14].

The current paper starts from the compartmental models, then describes a relevant individual-based model (i.e., the Eubank model [7]), and delves into a recent extension reported in [14].

2 Compartmental vs. Individual Models

Compartmental models divide the population into compartments (groups of subjects with homogeneous characteristics) and describe the variation of the number of subject that moves from one compartment to another through differential equations.

For instance, the SIR model uses the following compartments: (i) S: susceptible, (ii) I: infected, and (iii) R: recovered. Furthermore, let:

M.S. Mohamad et al. (Eds.): *7th International Conference on PACBB*, AISC 222, pp. 93–100.
DOI: 10.1007/978-3-319-00578-2_13   © Springer International Publishing Switzerland 2013

- β be the contact rate, i.e. the rate of becoming infectious by contacting another susceptible subject;
- γ be the recovery rate, i.e. the rate of recovering from an infection;

the variations in the time of the number of susceptible, infections and recovered individuals are thus described by the following equations:

$$\frac{dS}{dt} = -\beta IS; \quad \frac{dI}{dt} = \beta IS - \gamma I; \quad \frac{dR}{dt} = \gamma I \qquad (1)$$

According to (1), in the time, the number of susceptible individuals S decreases as of the infections (calculated in terms of the contact rate, number of susceptible and infected individuals), and the number of infected individuals I increases of the previous quantity and decreases as of the individuals that recovers from the infections (calculated in terms of the recovery rate and the number of infected individuals).

The SIR model can be extended by including, i.e. the birth/death rate, vaccinations, and by even adding further compartments, thus leading to more complex models (e.g. the SIER model).

It is worth stressing that the compartmental models are valid only in case of sufficiently large populations. Since a large population comprises of many different individuals in various fields, the diversity is reduced to a few key characteristics which are relevant to the infection under consideration, thus smoothing over the differences of each individual (e.g. specific behaviours, movements).

Instead, the so-called individual-based models try to take into account the specificities of the individuals composing the population and the way in which each individual can differently contract the disease or infect another individual. In particular, such approaches:

- build up a social network that realistically estimates the way in which every individual may contact other individuals;
- divide the epidemic process in terms of two sub-models, called between-host disease transmission and within-host disease progression. The first takes care of describing the disease transmission from one individual to another, the second takes care of describing the disease progression within each individual.

One of the most known individual-based models is the one proposed by Eubank at al [7]. Such a model has the following characteristics:

- The social network is build through a software agent called TRANSIM [13][3], that is able estimate the movements of each individual in a urban area;
- The between-host disease transmission model is based on bipartite graphs, in which the two classes of vertices are persons and locations, and the edges connects individuals to locations with a label that specifies the period of time in which an individual visited a location. For instance, the graph depicted in Fig. 1 shows person p_2 in L_1 from 8:00 to 9:00, in L_2 from 10:00 to 12:00, and in L_3 from 9:00 to 10:00;
- The within-host disease progression is modelled as follows: an individual becomes infected if in the same place of an infected individual for more than a certain period of time, and depending on the disease infectious rate.

It is worth noting that Eubank et al. estimated the social network for Portland (Oregon, USA), simulated a smallpox epidemics, and demonstrated – differently from the compartmental models that would have suggested mass vaccination – that the epidemics could have been better controlled through focused quarantine and vaccinations, combined with early detection.

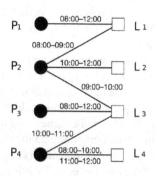

To understand such a result, let us take into account Fig. 2. Each node of the graph is an individual, and each edge between nodes represents that the two individuals are in contact each other, and thus may transmit each other an infectious disease. Let us suppose that the epidemic starts in an individual belonging to the area enclosed in the dashed box. For the epidemic to spread, it has to pass through the "bottleneck". Therefore, if we

Fig. 1. Example of bipartite graph estimated by TRANSIM

could have (i) an early detection system able to signal the presence of a possible outbreak and (ii) the social network of the population under analysis, we could stop the outbreak by putting into quarantine the sub-graph enclosed in the dotted box, and by targeted vaccination.

3   The Extended Model

The Eubank model uses exact movements of people (in order to properly label the arcs) and does not take into account the evolution of infectious diseases spread by vectors (the only classes of nodes are in fact people and places).

In this section, we summarize an extension of this model [14], that addresses the two limitations above as follows. In the between-host disease transmission model we introduce: (i) probability functions that captures the uncertainty about the movements of people and (ii) a further class of nodes representing vectors. Furthermore, we make use of probabilistic timed automata [5][6] to model the within-host progression.

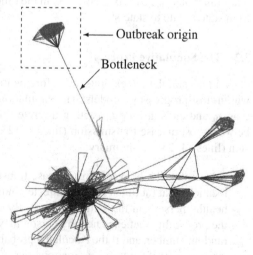

Fig. 2. Sample social network

3.1   The Mathematical Basis

The between-host model is a tripartite graph with three types of vertices that represent people, locations and vectors. A person is identified with the notation p, a vector with v, a place with

l. Let us assume respectively N, V and L be the number of individuals, vectors and places. Furthermore, at each discrete time t, a person p (or a vector v) is in only one place.

The edge that connects a person p (or a vector v) to a place l is labelled by a *probability function* $f_{p,l}(t)$ (or $f_{v,l}(t)$ for a vector), which represents the probability that, at time t, person p (or vector v) is in location l.

A person can contract the disease either by means of a contact with an infected person p' located in the same place of p, or due to the presence of a vector v. In addition, the individual may contract the disease or not according to a certain probability, which may depend on various factors (e.g. by the immune resources of the subject, the specificities of the disease, etc). Therefore, the probability that subject p becomes infected because of the infected individual p', in l, at time t, is given by:

$$f_{p,l,p'}(t) = \gamma_{p,p'}(t) \cdot f_{p,l}(t) \cdot f_{p',l}(t) \tag{2}$$

where $\gamma_{p,p'}(t)$ is the probability of disease transmission from p to p'.

Similarly, if the disease is transmitted by vectors, the probability for person p to contract the disease due to the presence of vector v in location l in time t is given by:

$$f_{p,l,v}(t) = \tau_{p,v}(t) \cdot f_{p,l}(t) \cdot f_{v,l}(t) \tag{3}$$

where $\tau_{p,v}(t)$ is the probability for person p to contract the disease from vector v. It is worth remarking that the model can take into account also aggregations of persons and vectors (e.g. a swarm of mosquito). Further details on this can be found in [14].

The within-host disease evolution is modelled as a finite state automata with probabilistic transitions [5], in a manner similar to that proposed in the work of Dodds & Watts [6]. The states of the automata represent the state of health of subject p (e.g., healthy, infected or dead), while the edges that connect the states are labelled by probability functions $f_{p,s,s'}(t)$ that describe, in the time, the probability for subject p to move from state s state to state s'.

3.2 The Simulative Process

An ad-hoc simulation exploits the aforementioned between-host transmission and within-host progression models. The simulation summarised in Algorithm 1 starts at time t_0 and ends at time t_1, with a discrete time interval Δt. It simulates firstly the between-host disease transmission (lines 4–12), then the within-host disease progression (lines 14–23). In summary:

- Concerning the between-host disease transmission, the algorithm cycles over all locations that (at time t) has an infected individuals on them. Then, cycles over all healthy persons in the same locations. By using equations (2) and (3), calculates the probability that each healthy person may become infected. Finally, it extracts a random number, and if the calculated probability is larger than the random number, we assume that the transmission took place;

- Concerning the within-host disease progression, the algorithm cycles over all non-healthy people and examines all possible state evolutions of the related probabilistic timed automata. Similarly, by using a random number, the disease evolves accordingly.

Algorithm 1. Iterative simulation

```
 1  t = t₀
 2  while  t ≤ t₁  do
 3
 4      //  BETWEEN–HOST DISEASE TRANSMISSION
 5      foreach location l with an infected person p' or vector v do
 6          foreach healthy person p with an edge in l do
 7              prob = f_{p,l,p'}(t) or p₂ = f_{p,l,v}(t)
 8              r = random [0,1)
 9              if (prob > r) then
10                  p becomes infected
11          end
12      end
13
14      //  WITHIN–HOST DISEASE PROGRESSION
15      foreach non–healthy person p do
16          let s be the state of p
17          foreach state s' connected to s do
18              prob = f_{p,s,s'}(t);
19              r = random [0,1)
20              if (prob > r) then
21                  p is in state s'
22          end
23      end
24
25      t = t + Δt
26  end while
```

It is clear that repeated executions of Algorithm 1, even on the same scenario, can produce different results, since both the between-host transmissions and the within-host progressions are influenced by the extraction of a random number. Therefore, similarly to the well-know Montecarlo simulations, we must repeat the algorithm as long as an adequate stop criterion is fulfilled. According to the central limit theorem, the algorithm is stopped when the following condition occurs:

$$2 \cdot x_{\alpha/2} \cdot \sqrt{\frac{S_N^2(t)}{N}} < \epsilon \qquad (4)$$

where N is the current number of iterations, $S_N^2(t)$ is the variance of the number of subjects belonging to the desired state (e.g. infected, dead), $1 - \alpha$ is the confidence level, $x_{\alpha/2}$ is chosen so that $\int_{-\infty}^{x_{\alpha/2}} g(t)dt = 1 - \alpha/2$, $g(t)$ is the normal distribution, and ϵ is the acceptable error level.

It is worth noting the impact of the stabilisation process. During the different iterations, different scenarios are generated. By comparing them, we could isolate the worst case (e.g. when we have the largest number of deaths), the best case, and the average one, so to have further opportunities to decide the best preventive and/or healing action.

4 Case Study

The case study describes the effect of probability in the transmission of the disease, in comparison with the model proposed by Eubank et al. In particular, let us refer to the following scenario. Let us suppose that an infected person visits a gym from 12:00 to 13:00, and 50 healthy individuals arrive at 13:00 and go off at 14:00, each for a different location. Furthermore, let us assume that the disease is such as to have an infectious rate of 10%.

In the model of Eubank et. al., the disease transmission occurs only if one can assume a contact for a more than a certain period of time. Therefore, in the above scenario, given the absence of any contact between the infected and the healthy individuals, the approach of Eubank at al. deduces the impossibility of the outbreak.

However, let us suppose that we could introduce uncertainty about the time in which the infected person visits the gym, and in particular let us suppose that we could model this uncertainty with a Binomial random variable with parameters $k = 11$ and $k \cdot q = 2$, as shown in Fig. 3. Accordingly, equations (2) and (3) become:

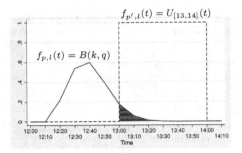

Fig. 3. Binomial $B(k,q)$ and uniform $U_{[13,14]}(t)$ variables

$$\left.\begin{array}{l} f_{p,l}(t) = B(k,q) \\ f_{p',l}(t) = U_{[13,14]}(t) \\ \gamma_{p,p'}(t) = 0.1 \\ \tau_{p,v}(t) = 0 \end{array}\right\} \Rightarrow \begin{array}{l} f_{p,l,p'}(t) = 0.1 \cdot B(k,q) \cdot U_{[13,14]}(t) \\ f_{p,l,v}(t) = 0 \end{array}$$

In such a context, it is now possible for the infected individual to contact the healthy persons and thus to give birth to an epidemic. In particular, the area behind the tail of the binomial distribution from 13:00 to 13:40 represents such a probability.

By using the simulation described above, we estimated that, at the end of the simulation ($t_0 = 12, t_1 = 14$):

- the best case is that the epidemic does not spread;
- the worst case is that we can have three infected people;
- the average case is to have more than one infected individual.

Fig. 4 shows the average number of infected in terms of probability of contact between the infected individual and healthy subjects (i.e., the tail of the Binomial distribution), and the infectious rate of the disease. As can be noticed, the simulations show that the average number of infected individuals grows as both the infectious rate and the probability of contact increase.

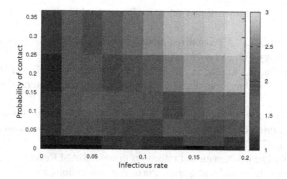

Fig. 4. Case study results (number of infections vs infectious rate and probability of contact)

5 Conclusions

The paper summarises few approaches that take into account disease outbreaks, starting with the traditional mathematical approaches (e.g. SIR) and ending with individual-based ones. The advantages of individual-based approaches were essentially connected with the ability of gaining better insights into the epidemics. Nevertheless, there are several issues that currently limit their application in real scenarios:

- The creation of the social network is an extremely complex process. Although e.g. a software as TRANSIMS can estimate the contact network, it is limited to only urban areas. However, the possibility of approximating portions of the population by adequately modifying the probability distributions is a possible approach [14];
- The execution of the simulation is an extremely expensive process from a computational point of view, and the resources required are very large. In this direction, there are proposals that attempt either to reduce the algorithmic complexity of the simulation [14] or to use parallel architectures [2].

References

1. Alloran, M.E., Longini Jr., I.M., Nizam, A., Yang, Y.: Containing bioterrorist smallpox. Science 298, 1428–1432 (2002)
2. Barrett, C.L., Bisset, K.R., Eubank, S.G., Feng, X., Marathe, M.V.: EpiSimdemics: an efficient algorithm for simulating the spread of infectious disease over large realistic social networks. In: Proceedings of the 2008 ACM/IEEE Conference on Supercomputing, SC 2008, pp. 1–12. IEEE Press, Piscataway (2008)
3. Beckman, R., Berkbigler, K., Bisset, K., Bush, B., Eubank, S., Henson, K., Hurford, J., Kubicek, D., Marathe, M., Romero, P., Smith, J., Smith, L.L., Speckman, P., Stretz, P., Thayer, G., Eeckhout, E.V., Barrett, C., Williams, M.: TRANSIMS, Los Alamos National Laboratory, ch. 3 (2003),
   http://ndssl.vbi.vt.edu/Publications/TransimsVol3Ch3.pdf
4. Carley, K., Fridsma, D., Casman, E., Yahja, A., Altman, N., Chen, L.C., Kaminsky, B., Nave, D.: BioWar: scalable agent-based model of bioattacks. IEEE Transactions on Systems, Man and Cybernetics, Part A: Systems and Humans 36(2), 252–265 (2006)

5. Danièle, B.: On probabilistic timed automata. Theoretical Computer Science 292(1), 65–84 (2003)
6. Dodds, P.S., Watts, D.J.: A generalized model of social and biological contagion. Journal of Theoretical Biology 232, 587–604 (2005)
7. Eubank, S., Guclu, H., Kumar, A., Marathe, M., Srinivasan, A., Toroczkal, Z., Wang, N.: Modelling disease outbreaks in realistic urban social networks. Nature 429(6988), 180–184 (2004)
8. Godfrey, K.: Compartmental models and their application. Academic Press (1983)
9. Grassly, N.C., Fraser, C.: Mathematical models of infectious disease transmission. Nat. Rev. Microbiol. 6(6), 477–487 (2008)
10. Hethcote, H.W.: The mathematics of infectious disease. SIAM Review 42, 599–653 (2000)
11. Keeling, M.: The effects of local spatial structure on epidemiological invasions. Proceedings of the Royal Society of London B Biological Sciences 266, 859–867 (1999)
12. Kenrad, N.E., Masters, C.F.: Infectious Disease Epidemiology: Theory and Practice. Jones & Bartlett Publishers (2006)
13. Rickert, M., Nagel, K.: Dynamic traffic assignment on parallel computers in TRANSIMS. Future Gener. Comput. Syst. 17(5), 637–648 (2001)
14. Vittorini, P., Villani, A., Di Orio, F.: An individual-based networked model with probabilistic relocation of people and vectors among locations for simulating the spread of infectious diseases. Journal of Biological Systems 18(04), 847–866 (2010)

A Statistical Comparison of SimTandem with State-of-the-Art Peptide Identification Tools[*]

Jiří Novák[1], Timo Sachsenberg[2], David Hoksza[1], Tomáš Skopal[1],
and Oliver Kohlbacher[2]

[1] Charles University in Prague, Faculty of Mathematics and Physics, Malostranské nám. 25,
118 00 Prague, Czech Republic
{novak,hoksza,skopal}@ksi.mff.cuni.cz
[2] Eberhard-Karls-Universität Tübingen, Applied Bioinformatics Group, Sand 14, 72076
Tübingen, Germany
{sachsenb,kohlbacher}@informatik.uni-tuebingen.de

Abstract. The similarity search in theoretical mass spectra generated from protein sequence databases is a widely accepted approach for identification of peptides from query mass spectra generated by shotgun proteomics. Since query spectra contain many inaccuracies and the sizes of databases grow rapidly in recent years, demands on more accurate mass spectra similarities and on the utilization of database indexing techniques are still desirable. We propose a statistical comparison of parameterized Hausdorff distance with freely available tools OMSSA, X!Tandem and with the cosine similarity. We show that a precursor mass filter in combination with a modification of previously proposed parameterized Hausdorff distance outperforms state-of-the-art tools in both – the speed of search and the number of identified peptide sequences (even though the q-value is only 0.001). Our method is implemented in the freely available application SimTandem which can be used in the framework TOPP based on OpenMS.

Keywords: peptide identification, tandem mass spectrometry, similarity search, parameterized Hausdorff distance, precursor mass filter, SimTandem.

1 Introduction

High performance liquid chromatography combined with tandem mass spectrometry (HPLC-MS/MS or shotgun proteomics) is a widely used technique for identification and quantification of proteins and peptides in complex mixtures. Mixtures obtained by a cell lysis contain thousands of proteins and a mass spectrometer produces tenths of thousands of peptide mass spectra (or query spectra) which must be annotated with peptide sequences [3].

Before a mass analysis, proteins in a sample are usually enzymatically digested to peptides. After chromatographic separation peptides are commonly subjected to electro spray ionization leading to positively charged ions. After transfer into the mass spectrometer, the most intense peptide ions are collected based on their mass-to-charge ($\frac{m}{z}$)

[*] This work was supported in part by FEBS Short-Term Fellowship, by Czech Science Foundation (GAČR) project Nr. 202/11/0968 and by the grant SVV-2013-267312.

M.S. Mohamad et al. (Eds.): 7th International Conference on PACBB, AISC 222, pp. 101–109.
DOI: 10.1007/978-3-319-00578-2_14   © Springer International Publishing Switzerland 2013

ratios and fragmented in a collision chamber. A list of $\frac{m}{z}$ ratios of fragment ions with intensities of their occurrence (i.e., a list of peaks) forms a tandem mass spectrum. The most common types of fragment ions occurring from collision induced dissociation techniques are y-ions and b-ions. Therefore, these ion types serve as main features for the annotation of spectra with peptide sequences. In practice, many peptides carry additional chemical modifications which change masses of amino acids, shift $\frac{m}{z}$ ratios of fragment ions and complicate the identification of peptide sequences [14].

The annotation of spectra with peptide sequences is often realized by means of a similarity search in databases of theoretical spectra generated from databases of known protein sequences, by the de-novo peptide sequencing, sequence-tag methods and comparison against a library of experimental spectra [11]. When the similarity search is utilized, protein sequences are algorithmically digested into shorter peptide sequences and theoretical peptide spectra are generated. Then spectra captured by a spectrometer (i.e., the query set) are compared with the theoretical spectra. Since databases of protein sequences grow rapidly in recent years, a comparison of all spectra in the query set with all theoretical spectra is time consuming. Fortunately, the search space of putative peptides can be greatly reduced by incorporating the precursor mass (mass of a peptide ion before fragmentation). In consequence, a query spectrum does not have to be compared with all theoretical spectra but only with a small subset of spectra in a precursor mass error tolerance λ.

A query spectrum is compared with theoretical spectra using a pair-wise similarity function and the nearest spectrum is selected. A peptide sequence corresponding to the nearest spectrum and the query spectrum form a peptide-spectrum match (PSM). Each PSM is accompanied by a score determined by the similarity function. A natural and common similarity score is the cosine similarity [8]. We proposed the parameterized Hausdorff distance which is able to identify more peptides than cosine similarity [12]. Tools based on the similarity search against databases of theoretical spectra like SEQUEST [4], MASCOT [13], OMSSA [5] and X!Tandem [2] implement their own similarity functions.

Even though different tools use different similarity functions, their performance can be compared by statistical evaluation [6]. A widely accepted technique is to apply a target-decoy approach. Protein sequences in a database are reversed and appended to the original database. Original sequences are marked as target sequences while reversed sequences are marked as decoy sequences. The false discovery rate can than be estimated as $FDR = \frac{\#decoy\ PSMs}{\#target\ PSMs}$. Since FDR is a property of a set of PSMs, the q-value is defined as minimum FDR threshold at which a given PSM is accepted as correct [6].

Let's assume a pair-wise distance function d. When a query set of spectra is compared against theoretical spectra, a set of PSMs is obtained where d_i is the distance between i^{th} spectrum in the query set and its nearest theoretical spectrum. Let t be a threshold of d and S be a set of PSMs such that $S = \{PSM_i \in S, d_i \leq t\}$. Now assume the following example. When $t = 0.6$ and S contains 5 decoy and 500 target PSMs, the $FDR = 0.01$. When $t = 0.65$ and S contains 5 decoy and 1000 target PSMs, the $FDR = 0.005$. Since the numbers of decoy PSMs are equal in both cases, the q-value is 0.005 for target PSMs.

2   Method

Here, we propose an approach for identification of peptides based on the similarity search of a query spectrum in a database of theoretical spectra. We describe the mass spectra similarity functions, the method how we speed-up the database search using the precursor mass filter and the method how we deal with modifications in spectra. The approach is implemented in the search engine SimTandem which can be easily used for batch analysis in TOPP (The OpenMS Proteomics Pipeline) [15] [7]. OpenMS is an open-source C++ library for LC-MS/MS data management and analyses which enables a statistical evaluation of results from different engines. Thus the engines can be easily compared. SimTandem is freely available at http://www.simtandem.org or http://www.siret.cz/simtandem.

2.1   Similarity Functions

When the similarity search in a database of theoretical spectra is employed for identification of peptides, a pair-wise similarity (or distance[1]) function is a crucial component of each search engine. The angle distance, the parameterized Hausdorff distance and a modification of the parameterized Hausdorff distance are defined below.

Angle Distance. The angle distance d_A (normalized dot product or cosine similarity) is a commonly utilized function for mass spectra comparison (Eq. 3) [8]. A representation of mass spectra as high-dimensional boolean vectors is usually used for this purpose. The range of $\frac{m}{z}$ values in a spectrum is split into subintervals. A width of a subinterval is determined by $\frac{m}{z}$ error tolerance ξ (e.g., $\xi = 0.5$ Da). When a peak falls into a subinterval, a boolean vector contains 1 at the position corresponding to the subinterval, otherwise it contains 0 (Fig. 1).

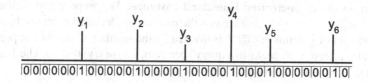

Fig. 1. High-dimensional boolean representation of a theoretical mass spectrum containing y-ions

Instead of storing high-dimensional sparse vectors, we use directly the vectors of $\frac{m}{z}$ values x and y (say, a low-dimensional representation of vectors). Considering the low-dimensional representation, two $\frac{m}{z}$ values between compared spectra are matched when $d_a(x_i, y_j) \leq \xi$. When the $\frac{m}{z}$ values are matched, the 1 is added to a sum. The max is used to prevent duplicate matches of the same $\frac{m}{z}$ value in one spectrum with more $\frac{m}{z}$ values in the other spectrum, i.e., every match of an $\frac{m}{z}$ value is counted only once.

---

[1] The smaller distance means the bigger similarity and vice versa.

$dim(x)$ is the dimension of x. Note that subintervals are not bounded as shown in Fig. 1 because the differences between $\frac{m}{z}$ values are computed.

$$d_a(x_i, y_j) = \begin{cases} 0, & \text{if } |x_i - y_j| > \xi \\ 1, & \text{else} \end{cases} \qquad (1)$$

$$a(x, y) = \sum_{x_i \in x} \max_{y_j \in y} \{ d_a(x_i, y_j) \} \qquad (2)$$

$$d_A(x, y) = \arccos \left(\frac{a(x, y)}{\sqrt{dim(x)dim(y)}} \right) \qquad (3)$$

Parameterized Hausdorff Distance. The parameterized Hausdorff distance d_{HP} (Eq. 6) [12] outperforms d_A in the number of identified peptides. The $\frac{m}{z}$ ratio y_j in the minimum distance $d_h(x_i, y_j)$ is found for each $\frac{m}{z}$ ratio x_i and then the nth root is used to reduce the impact of a big value in the sum (Eq. 5). The number of identified peptide sequences increases with increasing n [12]. For $n \to \infty$, the nth root converges to 1. Since lists of peaks in mass spectra are implicitly sorted, the d_{HP} is computed in linear time complexity $O(dim(x) + dim(y))$.

$$d_h(x_i, y_j) = \begin{cases} |x_i - y_j|, & \text{if } |x_i - y_j| > \xi \\ 0, & \text{else} \end{cases} \qquad (4)$$

$$h(x, y) = \frac{\sum_{x_i \in x} \sqrt[n]{\min_{y_j \in y} \{ d_h(x_i, y_j) \}}}{dim(x)} \qquad (5)$$

$$d_{HP}(x, y) = \max(h(x, y), h(y, x)) \qquad (6)$$

Modification of Parameterized Hausdorff Distance. We propose a modification of d_{HP} called d_{HP}^{match} (Eq. 8) which increases the number of identified peptides. In contrast to d_{HP}, the sum of $\frac{m}{z}$ ratios in d_{HP}^{match} is divided by the number of matches of peaks in a theoretical spectrum with peaks in a query spectrum, i.e., $a(x, y)$ (Eq. 2). The 1 is added to $a(x, y)$ to prevent the division by zero when $a(x, y) = 0$.

$$h^{match}(x, y) = \frac{\sum_{x_i \in x} \sqrt[n]{\min_{y_j \in y} \{ d_h(x_i, y_j) \}}}{dim(x)(a(x, y) + 1)} \qquad (7)$$

$$d_{HP}^{match}(x, y) = \max(h^{match}(x, y), h^{match}(y, x)) \qquad (8)$$

2.2 Precursor Mass Filter

Peptide precursor masses are known for both – theoretical and query mass spectra. Thus a query spectrum does not have to be compared with all theoretical spectra D generated from a database of protein sequences but only with a small subset $D_\lambda \subset D$ within a precursor mass error tolerance λ. For efficient determination of D_λ, D is sorted by precursor masses and D_λ is found by a binary search of the precursor mass of the query spectrum. Afterwards, theoretical spectra in D_λ are compared with the query spectrum using a distance function and the nearest theoretical spectrum to the query spectrum is selected. The query spectrum, the nearest theoretical spectrum and the score determined by the similarity function form a PSM.

2.3 Dealing with Modifications

Many peptides contain modifications which change masses of amino acids and thus shift $\frac{m}{z}$ ratios in the query spectra [14]. Modifications can be artificially added to a sample because they enable more precise analysis. They can arise during a sample preparation or during mass analysis. Post-translational modifications arise during the lifetime of a protein molecule and they give new properties to proteins, make stable conformations of proteins, regulate protein functions, etc. Modifications are fixed or variable. Fixed modifications change all amino acids of the same type, e.g., carbamidomethylation of cysteine. When a fixed modification is supported by a search engine, a mass of an amino acid is changed when theoretical spectra are generated, e.g., the mass of *cysteine* is increased by approx. 57.02 Da. Variable modifications do not have to change all amino acids of the same type, e.g., oxidation of methionine. When a variable modification is searched, theoretical spectra with all possible shifts of $\frac{m}{z}$ ratios are generated, compared with the query spectrum and the theoretical spectrum with the best score is selected to form a PSM.

3 Results

We used HPLC-MS/MS spectra from E. coli and human. Separation of the E. coli digest was performed using an easyLC HPLC system (Proxean) with a 2h segmented gradient. Peptides eluting from the column were online injected into an LTQ-Orbitrap XL instrument (Thermo Fisher Scientific), with top 10 selection of the most abundant ions for further fragmentation. A dynamic exclusion list of 500 masses and exclusion time of 90 seconds was used to avoid repeated fragmentation of the same ions. The query set *E.coli* contained 30,358 tandem mass spectra. Human spectra were taken from 2 runs from a label-free human data set [1] – the query set *Hum48* contained 26,417 spectra and *Hum49* contained 24,537 spectra. The data sets are available at http://www.simtandem.org.

The manually curated database containing 8,272 protein (332,862 peptide) sequences was used with *E.coli*. The database of 173,450 human protein (9,567,012 peptide) sequences from UniProtKB/Swiss-Prot (v. 07/2012) [16] was used with human query sets. Decoy protein sequences were included in both databases. Theoretical spectra were generated with following settings – enzyme: trypsin ([KR]/P); max. missed cleavage sites: 1; length of peptide sequences: 7-50 amino acids; precursor mass of peptides: 500-5,000 Da; fragment ions types: y, b, y^{2+}; $\frac{m}{z}$ ratios of fragment ions: 200-2,000 Da. Query spectra were processed as follows – minimum number of peaks in a spectrum to be processed: 30; peak selection heuristic: the range of $\frac{m}{z}$ values was split by 50 Da, 5 most intense peaks were selected in each window and 50 most intense peaks were

Table 1. Numbers of identified peptides and search times [min:sec]. In a cell with a number of identified peptides, the best result among all engines is highlighted.

		OMSSA				X!Tandem			
		q-value			Time	q-value			Time
		0.05	0.01	0.001		0.05	0.01	0.001	
E.coli		11,729	10,301	7,989	03:40	10,277	8,518	6,398	03:35
	mod.	13,008	11,435	9,123	05:00	11,612	9,813	7,487	04:36
Hum48		6,893	6,147	**5,439**	27:42	7,330	5,902	4,239	40:03
	mod.	9,430	8,508	**7,229**	27:42	**10,494**	8,524	5,411	51:20
Hum49		8,118	7,119	**5,392**	24:03	7,728	6,085	4,673	37:07
	mod.	11,465	10,333	8,601	25:10	11,695	9,712	6,119	53:30

		d_A				d_{HP}				d_{HP}^{match}			
		q-value			Time	q-value			Time	q-value			Time
		0.05	0.01	0.001		0.05	0.01	0.001		0.05	0.01	0.001	
E.coli		12,846	10,404	1,785	00:40	13,004	11,200	8,307	00:44	**13,373**	**11,668**	**9,402**	00:43
	mod.	14,554	11,587	1,948	01:26	14,576	12,556	9,288	01:33	**14,969**	**13,113**	**10,005**	01:30
Hum48		6,205	3,717	773	03:19	7,162	5,863	4,225	03:53	**7,461**	**6,230**	4,497	03:42
	mod.	8,615	4,845	934	06:06	9,882	8,108	6,119	06:29	10,305	**8,680**	7,109	06:34
Hum49		7,859	5,186	2,512	03:05	8,347	6,888	5,329	03:44	**8,760**	**7,512**	5,311	03:57
	mod.	11,244	6,944	3,291	05:12	12,002	10,011	8,120	06:46	**12,531**	**10,816**	**8,689**	06:15

selected from the unification of the most intense peaks in the windows. $\lambda = 10$ ppm, $\xi = 0.5$ Da and $n = 30$ (in d_{HP} and d_{HP}^{match}). A machine with Windows 7 x64, Intel Core i7 2GHz, 8 GB RAM and 5400 rpm HDD was used.

3.1   State-of-the-Art Tools

Numbers of identified peptides for different q-values and search times were measured for freely available tools OMSSA (v. 2.1.8) and X!Tandem (v. 2011.12.01.1). The refinement mode in X!Tandem was not used because it impacted the statistical evaluation. The comparison was made using OpenMS (v. 1.9). Simple pipelines in TOPPAS were created for this purpose (e.g., *OMSSAAdapter* → *PeptideIndexer* → *FalseDiscoveryRate* → *IDFilter*). Pipelines were processed without and with the support of modifications (*carbamidomethylation of cysteine* was used as a fixed modification and *oxidation of methionine* as a variable modification). Results are shown in Tab. 1. OMSSA identified more peptides than X!Tandem in all query sets and the search was $1.5\times$-$2.1\times$ faster on human query sets.

3.2   SimTandem

Numbers of peptides identified by SimTandem (i.e., by precursor filter with d_A, d_{HP} and d_{HP}^{match}) and search times are shown in Tab. 1. The most peptides are identified when d_{HP}^{match} is used. However, when $q = 0.001$, OMSSA identifies more peptides in three cases. When $q = 0.05$, X!Tandem identifies the most peptides in one case. d_{HP}^{match} identifies more peptide sequences than d_{HP} in almost all cases. The number of identified peptides is bigger for d_{HP} than for X!Tandem when E.coli and Hum49 are used. OMSSA outperforms d_{HP} on human query sets when $q \le 0.01$ is used. The number of identified peptides is significantly smaller for d_A than for other engines.

SimTandem (precursor filter with d_{HP}) is $5.0\times$-$7.1\times$ faster than OMSSA without the support of modifications and $3.2\times$-$4.3\times$ faster with the support of modifications. It is also $4.9\times$-$10.3\times$ faster than X!Tandem without modifications and $3.0\times$-$7.9\times$ faster with modifications. The speed up is almost the same when d_{HP}^{match} is used.

A graphical comparison of numbers of identified peptides with state-of-the-art tools is shown in Fig. 2. The overlap of identified peptides between SimTandem and OMSSA is bigger than the overlap between SimTandem and X!Tandem and also bigger than the overlap between OMSSA and X!Tandem on all query sets.

We have tested the impact of the index of the root n in d_{HP} and d_{HP}^{match} on the number of identified peptides. Results are shown in Tab. 2. The number is bigger with bigger n. However, when n is too big, the number of identified peptides is smaller. The optimal n depends on the data sets and should be determined empirically.

3.3   Efficiency of Precursor Mass Filter

Since the number of comparisons of a query spectrum with theoretical spectra is crucial for the efficiency of the precursor mass filter, an average number of comparisons was measured in protein sequence databases Swiss-Prot (v. 07/2012) (human sequences only and all sequences) [16], MSDB [9] and NCBI RefSeq (v. 55) [10]. The query set *Hum48* was used and modifications were not supported. The results are shown in Tab. 3. Since an organism is usually known for a query set of spectra (e.g., E. coli or human)

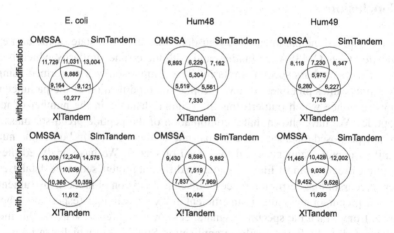

Fig. 2. Comparison of SimTandem (d_{HP}) with state-of-the-art tools ($q = 0.05$)

Table 2. Numbers of identified peptides by d_{HP} and d_{HP}^{match} for different n ($q = 0.001$, modifications in query spectra were not supported). The best result in each column is highlighted.

n	d_{HP}			d_{HP}^{match}		
	E.coli	Hum48	Hum49	E.coli	Hum48	Hum49
5	4,237	1,605	2,321	8,307	3,283	4,646
10	6,288	2,590	4,167	9,165	4,194	**5,517**
20	8,101	3,773	5,160	9,360	4,440	5,220
30	8,307	4,225	5,329	9,402	4,497	5,311
50	8,370	4,343	5,144	9,427	4,530	5,268
100	**8,415**	**4,447**	**5,579**	**9,453**	**4,560**	5,295
∞	8,348	3,928	5,023	9,332	3,988	5,110

Table 3. Average numbers of comparisons of a query spectrum with theoretical spectra for different protein sequence databases and different precursor mass error tolerances λ. Numbers of protein and peptide sequences in databases are also proposed (decoy sequences are included in the databases).

Database	Number of proteins	Number of peptides	λ					
			5 ppm	10 ppm	15 ppm	0.5 Da	1 Da	2 Da
Swiss-Prot (human)	173,450	9,567,012	206	409	613	3,892	7,792	15,564
Swiss-Prot (complete)	1,073,578	52,361,610	1,056	2,091	3,135	21,261	42,645	85,141
MSDB	6,478,158	281,767,270	5,756	11,369	17,042	113,272	227,017	453,153
NCBI	34,737,538	1,533,987,691	30,606	60,638	91,004	612,225	1,227,339	2,451,235

and the precision of modern spectrometers increases, the number of spectra compared with a query spectrum is small and thus the precursor filter is efficient. For example, 409 spectra are compared with a query spectrum when human sequences from Swiss-Prot are used and when $\lambda = 10$ ppm. When the NCBI database is used, the number of comparisons is 60,638. For a mass spectrometer with a low precision $\lambda = 2$ Da, the number of comparisons is significantly bigger. For example, 15,564 spectra are compared when human sequences from Swiss-Prot are used and 2,451,235 comparisons are made when the NCBI database is used.

4   Conclusions

We have compared the parameterized Hausdorff distance with state-of-the-art peptide identification tools OMSSA, X!Tandem and with the cosine similarity which is commonly utilized as a mass spectra similarity. The comparison was performed using the OpenMS proteomics pipeline. We have proposed a modification of the parameterized Hausdorff distance which outperforms the original distance in the number of identified peptides. We have shown that a combination of the peptide precursor mass filter with newly proposed distance outperforms state-of-the-art tools in both – the number of identified peptide sequences and the speed of search. We have analyzed the efficiency of the precursor mass filter considering different protein sequence databases and different precursor mass error tolerances. Since the precision of modern instruments increases, the precursor mass filter is an efficient indexing technique (especially, when the organism, from which the spectra originate, is known before the search). Our method is implemented in the freely available application SimTandem which can be used for batch processing using TOPP.

References

1. Beck, M., et al.: The quantitative proteome of a human cell line. Molecular Systems Biology 7, 549 (2011)
2. Craig, R., Beavis, R.C.: TANDEM: matching proteins with tandem mass spectra. Bioinformatics 20(9), 1466–1467 (2004)
3. Eidhammer, I., Flikka, K., Martens, L., Mikalsen, S.O.: Computational Methods for Mass Spectrometry Proteomics. John Wiley & Sons, England (2007)

4. Eng, J., McCormack, A., Yates, J.: An approach to correlate tandem mass spectral data of peptides with amino acid sequences in a protein database. J. of the Am. Soc. for Mass Spec. 5, 976–989 (1994)
5. Geer, L.Y., et al.: Open Mass Spectrometry Search Algorithm. Journal of Proteome Research 3, 958–964 (2004)
6. Käll, L., et al.: Assigning Significance to Peptides Identified by Tandem Mass Spectrometry Using Decoy Databases. Journal of Proteome Research 7, 29–34 (2008)
7. Kohlbacher, O., et al.: TOPP – the OpenMS proteomics pipeline. Bioinformatics 23(2), e191–e197 (2007)
8. Liu, J., et al.: Methods for peptide identification by spectral comparison. Proteome Science 5(3) (2007)
9. MSDB, http://www.proteomics.leeds.ac.uk/bioinf/
10. NCBI RefSeq, http://www.ncbi.nlm.nih.gov/RefSeq/
11. Nesvizhskii, A.I.: A survey of computational methods and error rate estimation procedures for peptide and protein identification in shotgun proteomics. Journal of Proteomics 73(11), 2092–2123 (2010)
12. Novák, J., Hoksza, D.: Parametrised Hausdorff Distance as a Non-Metric Similarity Model for Tandem Mass Spectrometry. In: CEUR Proc. DATESO, pp. 1–12 (2010)
13. Perkins, D.N., et al.: Probability-based protein identification by searching sequence databases using mass spectrometry data. Electrophoresis 20(18), 3551–3567 (1999)
14. Pevzner, P.A., et al.: Efficiency of Database Search for Identification of Mutated and Modified Proteins via Mass Spectrometry. Genome Research 11(2), 290–299 (2001)
15. Sturm, M., et al.: OpenMS – An open-source software framework for mass spectrometry. BMC Bioinformatics 9, 163 (2008)
16. UniProtKB/Swiss-Prot, http://www.uniprot.org/

HaptreeBuilder:
Web Generation and Visualization
of Risk Haplotype Trees

Dimitra Kamari[1], María Mar Abad-Grau[2,*], and Fuencisla Matesanz[3]

[1] Department of Computer Engineering and Informatics - University of Patras,
Patras, Greece
kamari@ceid.upatras.gr

[2] Department of Computer Languages and Systems - CITIC - University of Granada,
Granada, Spain
mabad@ugr.es

[3] Instituto de Parasitología López Neyra, CSIC, Granada, Spain
lindo@ipb.csic.es

Abstract. The quantity and quality of genome-wide association studies for several diseases are constantly increasing. As a consequence, molecular biologists from different laboratories demand new visualization tools for them to explore results by view and formulate new conjectures to work on. Although nowadays most studies are not able to reconstruct individual haplotypes, the next generation sequencying technologies will allow to obtain individual haplotypes in most studies conducted in the next few years. As evolutionary analysis of the haplotypes can be an invaluable information to biomedical researchers to build hypotheses of genetic variation by considering haplotype evolution, we have build a web-based tool for biomedical researchers to build and visualize risk haplotype trees along a chromosome so that they can perform a visual online exploration of the genetic factors associated with complex diseases.

Keywords: genome-wide association study, haplotype, disease risk, next generation sequencing.

1   Introduction

Although risk variants increasing individual predisposition to develop a complex disease usually require to know individual haplotypes, most of the current genome-wide association studies (GWAS) ignore haplotype configurations and only focus on genotype data. This simplistic approach still has helped to detect numerous risk variants for several complex diseases such as Multiple Sclerosis (MS) [1]. However, the most important results are expected to be obtained whenever the technology is mature enough for the molecular biologists to conduct GWAS using haplotypes instead of genotypes. Nowadays, to conduct a GWAS

---

* Corresponding author.

M.S. Mohamad et al. (Eds.): *7th International Conference on PACBB*, AISC 222, pp. 111–117.
DOI: 10.1007/978-3-319-00578-2_15   © Springer International Publishing Switzerland 2013

using haplotypes is very expensive and very few have been performed. As examples of diseases for which one of these studies exists are MS, dental caries and attention deficit hyperactivity disorder. For most studies, haplotypes may be inferred from genotypes by using computational algorithms. However, these algorithms are accurate only for haplotypes with a small number of markers [2]. Some association tests have been proposed in order to analyse whether association exists between haplotypes and a trait or disease. As their power is directly related with the accuracy of inferred haplotypes, they were succesful only when used with well-inferred haplotypes. It has to be noted, that the longer the haplotypes the larger the difficulty to be inferred without errors. Among them, we focus in this work in those able to use haplotype trees, i.e., ancestral relationships between haplotypes, to improve power when assessing association evidence of risk to develop a complex disease. Some examples are TreeDT [3], which infers a tree of haplotype ancestors given a data set of current haplotypes, TreeDTh [4] which improves TreeDT by keeping Type-I errors due to multiple testing under control while increasing power and 2GTree [5] which modifies 2G [6] by building trees using a Bayesian measure. As far as we know, neither of them provide a visual tool so that trees along a chromosome can be explored by a biomedical researcher. Moreover, results from every haplotype-based association test such as 2G are susceptible to be augmented with haplotype trees.

In this work we present HaptreeBuilder, a web-based and visual application whose main goal is to facilitate biomedical researchers to access to results from GWAS conducted on haplotypes. Together with association maps already known [6], the application is able to automatically build webpages for a given trait in which haplotype trees have been learned from data and draw to provide a graphical image to biomedical researchers. Section 2 describes HaptreeBuilder, including its main functional features, its architecture and the user interface. It also provides details about the more complex tasks and design and implementation issues. In Section 3 we show how the application has been used to build web pages showing haplotype-based results from a GWAS on MS. A discussion appears in Section 4. The website has been created at http://bios.ugr.es/haptreeBuilder. Supplementary material is accesible at http://bios.ugr.es/haptreeBuilder/supplementary.pdf.

2   Method

HaptreeBuilder is a web-based and visual tool with the main purpose of summarizing haplotype information and their associations with a trait by building risk haplotype trees from genome-wide association results and creating and displaying images of them. It is accesible at http://bios.ugr.es/haptreeBuilder.

2.1   Risk Haplotype Trees

A haplotype tree is a tree that has haplotypes as nodes. Several algorithms use haplotype trees. They usually consider the root node as the oldest haplotype and arcs are used to represent haplotype evolution. The length of the arc may be used to represent the time the new haplotype took to evolve. For instance, two nodes may be linked when the one closer to the root evolved to the other one by a mutation. Figure 1 (a) represents a haplotype tree with haplotypes composed of four nucleotides. Instead of using arrows for arcs, we used a rectangle to represent the root node. When two mutations were required to produce a new haplotype, as in the case of AAAA→CGAAA, the arc has double length than when only one mutation explains the new haplotype.

But HaptreeBuilder produces haplotype trees enriched with risk information. Therefore, as they are built in relation with a trait or a disease, each haplotype node can have a different color depending on whether it is a risk haplotype (red colour) or not (blue color). Figure 1 (b) shows an example of a haplotype tree with length of 10 nucleotides. To save space, haplotypes are represented by their codes in the plot. The actual haplotype configuration for this risk haplotype list can be seen in Table S1.

To build a haplotype tree from a list of haplotypes, we have used the TCS algorithm [7], an algorithm able to infer population level genealogies given a data set with the variants of a current population. It first computes a distance matrix with distances among every pair of haplotypes in the data set and then tries to connect the variants (haplotypes) by using a tree (cladogram). Table S2 gives more details about the algorithm.

Together with risk haplotype trees, HaptreeBuilder also provides risk maps, i. e. an horizontal bar for each locus whose length is proportional to the association p-value found at that locus given a trait.

2.2 Main Features

Users can access to HaptreeBuilder through four different roles: visitor, standard user, expert user and administrator. There is a common inheritance relationship among all these users (see Figure S1), being the visitor at the root, i.e. they will have access to very few functions, and the administrator at the bottom, i.e. they will have full access to the system features.

As the tool can be used by different user roles, we will describe the main features considering the more ancestral user role with access to each one:

The main feature of HaptreeBuilder are the following:

- Visitor: General information about the application
- Visitor: User registration and login
- Standard user: Visualization of risk maps
- Standard user: Visualization of risk haplotype trees
- Expert user: Produces a web site with all risk maps and haplotype trees for all the chromosomes given a trait and haplotype length.
- Administrator: Promotes a standard user to expert user or steps down an expert user to standard user. For security reasons, a user cannot reach the role of an expert user without the administrator consentment.

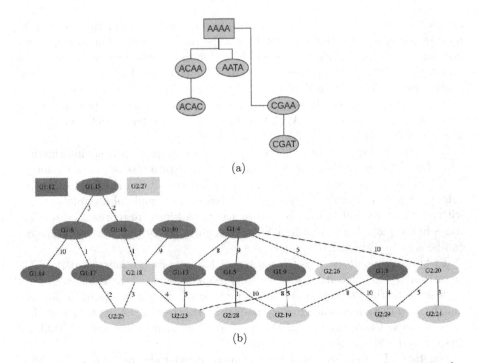

Fig. 1. (a) An example of a haplotype tree. The root node is represented as a rectangle. There may be more than one tree for a given locus. The length of an arc is proportional to the number of mutations required for an haplotype to evolve from its most recent ancestor. (b) An example of some risk haplotype trees built by HaptreeBuilder at a given locus. Nodes in red represent risk haplotypes, i.e. those found in association with the trait analyzed, while nodes in blue represent those haplotypes with no relation with the trait.

As the generation of the web site for a trait is quite complex and requires to use different software tools and data formats, we provide in Table S3 a detailed description of all the steps involved, the input data they require and the output. The user interface is a common graphical web-based interface. Figure 2 (a) shows the home page of the site.

2.3 System Architecture and Implementation Details

HaptreeBuilder has a client-server three-layer architecture having the server side one thick layer with controlling responsibilities and the most computationally demanding functionality, i.e. algorithms to build haplotype trees and maps, implemented under it. It also has a more basic layer with a data base managing system (DBMS) to control user credentials. The client side has only presentation/visualization reponsibilities. Figure 2 (b) shows the system architecture of HaptreeBuilder. HaptreeBuilder has been developed by using different software tools. To make the haplotype trees, a java implementation of TCS has been used

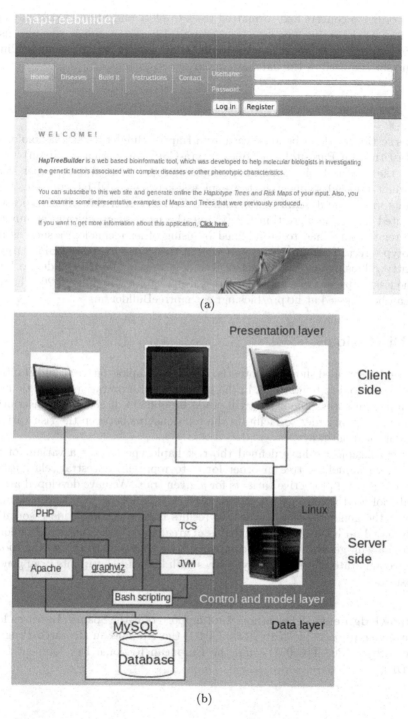

(a)

(b)

Fig. 2. (a) Home page of HaptreeBuilder. (b) System architecture of HaptreeBuilder.

[8]. In order to convert text-formatted trees to plots and to build risk maps we used php as the script language. The graphical library graphviz was also used to generate the plots. Web sites are built on an Apache webserver under linux and mysql was used as the database management system.

3   Use Case

In this section we describe a use case from HaptreeBuilder. Molecular biologists at "Instituto de Parasitología López Neyra-CSIC" in Granada (Spain) working in the genetic basis of multiple scleroris were interested in a deeper study of variant rs1299934, affecting the expresssion of gene DRB1, and the other variants in linkage disequilibrium with it. Association of these haplotypes was computed using the algorithm G2 [5] but the need to graphically summarize these relationships had to be fullfilled by using other technologies such as risk haplotype trees. Therefore, an expert user at the research group of Bioinformatics (http://bios.ugr.es) built a set of web pages using HaptreeBuilder to build all the risk maps and risk haplotype trees using haplotypes of legth 20. The web site can be accessed at http://bios.ugr.es/haptreeBuilder/ms.

4   Discussion

The results obtained show the potential of the risk haplotype trees, and Haptree-Builder as a tool to build them, in the search and interpretation of genetic basis of complex diseases. The same trees could be built in discretized transcription phenotypes [9] in order to facilitate the relationships between the risk variants and biological pathways.

As conclusion, we have defined the risk haplotype tree as a variant of the well-known haplotype tree in order for it to represent ancestral relationships between risk and protective variants for a given trait. We have developed a web-based tool for the automatic generation of risk haplotype trees and risk maps all over the genome given a trait and results from a genome-wide association study based on haplotypes. This tool is of great interest for molecular biologist researching on genetic basis of a complex disease. We plan to extend the tool in order to annotate the risk haplotype trees with known risk variants from genetic repositories.

Acknowledgments. The authors were supported by the Spanish Research Program under project TIN2010-20900-C04-1, the Andalusian Research Program under project P08-TIC-03717 and the European Regional Development Fund (ERDF).

References

1. (IMSGC), I.M.S.G.C.: Evidence for polygenic susceptibility to multiple sclerosis - the shape of things to come. Am. J. Hum. Genet. 86, 621–625 (2010)
2. Higasa, K., Kukita, Y., Kato, K., Wake, N., Tashira, T., Hayashi, K.: Evaluation of haplotype inference using definitive haplotype data obtained from complete hydatidiform moles, and its significance for the analyses of positively selected regions. PLoS Genetics 5, e1000468 (2009)
3. Sevon, P., Toivonen, H., Ollikainen, V.: Tree pattern mining for gene mapping. IEEE Transaction on Computation Biology and Bioinformatics 3(2), 174–185 (2001)
4. Moreno-Ortega, J.J., Medina-Medina, N., Montes-Soldado, R., Abad-Grau, M.M.: Improving reproducibility on tree based multimarker methods: TreeDTh. In: Rocha, M.P., Corchado Rodríguez, J.M., Fdez-Riverola, F., Valencia, A. (eds.) PACBB 2011. AISC, vol. 93, pp. 389–396. Springer, Heidelberg (2011)
5. Abad-Grau, M., Medina-Medina, N., Moral, S., Montes-Soldado, R., Torres-Sánchez, S., Matesanz, F.: Increasing power by using haplotype similarity in a multimarker transmission/disequilibrium test. Journal of Bioinformatics and Computational Biology 11(2), 1250014 (2013)
6. Abad-Grau, M., Medina-Medina, N., Montes-Soldado, R., Matesanz, F., Bafna, V.: Sample reproducibility of genetic association using different multimarker tdts in genome-wide association studies: Characterization and a new approach. PLoS One 7(2), 29613 (2012)
7. Templeton, A.R., Crandall, K.A., Sing, C.F.: A cladistic analysis of phenotypic associations with haplotypes inferred from restriction endonuclease mapping and DNA sequence data. III. cladogram estimation. Genetics 132(2), 619–633 (1992)
8. Clement, M., Posada, D., Crandall, K.: Tcs: a computer program to estimate gene genealogies. Molecular Ecology 9(10), 1657–1660 (2000)
9. Masegosa, A., Abad-Grau, M.M., Moral, S., Matesanz, F.: Learning classiers from discretized expression quantitative trait loci. In: IWBBIO 2013: Proceedings of the 1st International Work-Conference on Bioinformatics and Biomedical Engineering, Granada, Spain, pp. 1–8 (2013)

Speeding Up Phylogenetic Model Checking

José Ignacio Requeno and José Manuel Colom

Department of Computer Science and Systems Engineering (DIIS), Universidad de Zaragoza,
C/ María de Luna 1, 50018 Zaragoza, Spain
{nrequeno,jm}@unizar.es

Abstract. Model checking is a generic and formal technique that the authors have proposed for the study of properties that emerge from the biological labeling of the states defined over the phylogenetic tree [3] [10]. This strategy allows us to use generic software tools already present in the industry. However, the performance of traditional model checking is penalized when scaling the system for large phylogenies. To this end, two strategies are presented here. The first one consists of partitioning the phylogenetic tree into a set of related subproblems so as to speed up the computation time and distribute the memory consumption. The second strategy is based on uncoupling the information associated to each state of the phylogenetic tree (mainly, the DNA sequence) and exporting it to an external tool for the management of large information systems. The integration of all these approaches outperformed the results of monolithic model checking and helped us to execute the verification of properties in a real phylogenetic tree.

Keywords: distributed and sliced model checking, database, phylogenetic property.

1 Introduction

A phylogenetic tree is a description of the evolution process which is discovered through molecular sequencing data and morphological data matrices [8]. We applied model checking techniques in our previous works in order to study properties that emerge from the biological labeling of the tree states [3] [10]. For example, the next formula specifies the backmutations present in a branch of the phylogeny [10]: $BM\,(col,\sigma) \equiv (seq[col] = \sigma) \wedge \mathbf{EF}\,[(seq[col] \neq \sigma) \wedge \mathbf{EF}\,(seq[col] = \sigma)]$.

Model checking is a generic unifying formalism that allows the phylogeneticist to focus on tree structures, biological properties and symbolic manipulation of phylogenies described using temporal logic, instead of on implementation issues concerned with particular algorithms. Model checking allows us to uncouple software tools from the definition of properties and it hides the underlying implementation technology. Besides, these properties can be exported and evaluated in other structures (i.e., trees or networks) so as to compare the results and define a metric.

Nevertheless, the performance is penalized when scaling the system for large phylogenies and alignments [10]. The underlying problem of standard model checking is the great amount of phylogenetic data it has to deal with: the information associated to each node of the tree is strongly related to the DNA sequence of the specie (up to millions of nucleotides).

M.S. Mohamad et al. (Eds.): *7th International Conference on PACBB*, AISC 222, pp. 119–126.
DOI: 10.1007/978-3-319-00578-2_16 © Springer International Publishing Switzerland 2013

In order to solve this, two strategies are presented here. The first one consists of partitioning the global graph structure into a set of related subproblems so as to speed up the computation time and distribute the memory consumption. Two subtactics are considered here depending on the division method: the partition of the tree into subtrees, each one managed by a different model checker; or the slicing of the tree, each slice containing a copy of the original tree but only a portion of the DNA sequence. These techniques were presented in previous works [6] [5] [11], but we compare here the performance of these two complementary tactics.

The second strategy is based on uncoupling the DNA information of the Kripke structure and exporting the alignment to an external tool specialized in the management of large information systems, for example, a database.

In this paper, we show that a combination of both strategies allows us to obtain the best of the two worlds. Hence, this document is thus divided in 5 sections. After this introduction, Section 2 introduces the adaptation and implementation of distributed model checking techniques for the particular case of phylogenetic trees. Secondly, Section 3 presents the notions of external databases for storing the DNA alignment. Next, Section 4 summarizes how all these solutions can be integrated in a workflow that improves the performance of each solution in isolation. Finally, Section 5 briefs the conclusions.

2   Partitioned Model Checking

2.1   Division in Subtrees

Model checking based on distributed Kripke (tree) structures attempts to improve the performance by partitioning the graph into smaller subgraphs and distributing the chunks among available computing units (both the storage of the partial Kripke structure and computation of satisfiability of logic formulas [6]). These methods attack the size of the structure (number of states) and not the complexity of each state, which is the other limiting factor in phylogenetic model checking.

In the particular case of trees, the verification and communication process between chunks is simplified due to their inherent acyclicness. Given the tree root, s_0, we verify the property for each of the direct subtrees and operate with the boolean results according to the logical quantifiers. In the case of a generic formula ϕ written in "Computational Tree Logic" (CTL) [1], the equivalence is supported by the following recursive expansion law of the path operators (**EX, EG, E_U_**):

$$M, s_0 \models \mathbf{EX}\phi \equiv \bigvee_{s_i \in successors(s_0)} M_i, s_i \models \phi$$
$$M, s_0 \models \mathbf{EG}\phi \equiv M_0, s_0 \models \phi \wedge \left(\bigvee_{s_i \in successors(s_0)} M_i, s_i \models \mathbf{EG}\phi\right)$$

$$M, s_0 \models \mathbf{E}(\phi_1 \mathbf{U} \phi_2) \equiv M_0, s_0 \models \phi_2 \vee$$
$$(M_0, s_0 \models \phi_1 \wedge (\bigvee_{s_i \in successors(s_0)} M_i, s_i \models \mathbf{E}(\phi_1 \mathbf{U} \phi_2)))$$

where M is the original tree with root s_0. M_i is the subtree with root $s_i \in successors(s_0)$. The degenerated tree M_0 only needs the node s_0. The formula can be unfolded indefinitely by means of **EX**. In this case, the set of successors s_i defines a border at a certain depth of the original tree: we must ensure that ϕ holds in s_0, in the subtree rooted by s_i and in a path between s_0 and s_i. The root s_i acts as an interface node for the communication process during the composition of the partial results.

The number and size of the subtrees, and the appearance of short circuits during the composition of results will determine the performance of this approach. The detection of clades with conserved regions or characteristic SNP's are the kind of properties that benefit of this approach.

The implementation of the model checker is also important. For example, Table 1 represents the time needed for the initialization of the NuSMV model checker given a phylogenetic tree labeled with simple identifiers (GenBank accession numbers [2]). If the tree is balanced, the initialization and verification of the subtrees is faster than the original one because the trend is quadratic with respect to the tree size.

Table 1.

Nodes	165	435	631	785	1016	1365	1697	2136	5280	7602	13141	13444	14512
Seconds	0,012	0,060	0,112	0,156	0,240	0,444	0,668	1,012	7,876	15,361	48,679	52,643	57,820

2.2 Slicing the State

Phylogenetic model checking usually associates a complete DNA sequence per node of the tree. Sometimes the storage in local memory of the phylogenetic tree together with the atomic propositions leads to a high memory consumption in the model checker tool. In order to solve it, the state slicing (also called sliced model checking) focuses on the state complexity of the Kripke structure by creating several copies of the original phylogenetic tree and verifying the subproperties in parallel, each slice labeled with a partial substring of the DNA [11].

Consider a generic phylogenetic property $\phi(p_1, p_2, \ldots, p_l)$ represented by a temporal logic, with $p_i \in AP_i = \{seq[i] = \sigma | \sigma \in \Sigma\}$ an atomic proposition that asks for the tree nodes having the nucleotide σ in the i-th position of the DNA sequence (Def. 3, [11]). The classical model checking algorithm parses the formula and starts verifying the inner subformulas first. In the derivation tree from the CTL grammar for the given formula, we reach a propositional operator at some point of the recursion (boxed lines in Algorithm 1).

The verification of formulas with propositional operators, such as $\psi_1 \vee \psi_2$, begins with the computation of the satisfiability sets $Sat(M, \psi_j)$, $j = \{1, 2\}$. These sets allow us to distribute the computation in parallel, as long as we compose the partial results with a synchronized union. The support of ψ_j is $\|\psi_j\| = \{p_i \in AP_i | p_i \text{ or } \neg p_i \text{ appears in } \psi_j\}$. The verification of ψ_j is mapped to a remote model checking tool that has a copy of the phylogenetic tree labeled with the slice $R_I = \bigcup_{i \in I} AP_i$, with $I \subseteq \{1 \ldots l\}$ a set of index, $l = length(DNA)$ and $AP = \bigcup_{i=1}^{l} AP_i$. The remote model checker also executes the Algorithm 1 but with the set of atomic propositions belonging to $\|\psi_j\| \subseteq R_{I_j}$. In order to obtain a perfect distribution, $\bigcap_{j=\{1,2\}} \|\psi_j\| = \emptyset$. The notation $par(R_{I_j}, Sat(M, \psi_j))$ means that $Sat(M, \psi_j)$ is computed in parallel in the remote model checker R_{I_j} associated to the slice containing $\|\psi_j\|$. L_{R_I} is the labeling function of the states of that slice. The computation of the CTL paths is denoted with the fixpoint algorithms $\mu Z / \upsilon Z$.

Algorithm 1.. Algorithm $Sat\,(M,\phi)$

Require: $M = (S, S_0, R, L)$ is a Kripke structure; $\phi(p_1, p_2, \ldots, p_l)$ is a CTL formula

if $\phi \equiv \top$ **return** S {Set of states from the Kripke structure M}

else if $\boxed{\phi \equiv p_i \in AP}$ **return** $par(R_I, \{s : p_i \in L_{R_I}(s)\})$ with $i \in I$

else if $\boxed{\phi \equiv \neg \psi}$ **return** $S \setminus par(R_I, Sat(M, \psi))$

else if $\boxed{\phi \equiv \psi_1 \vee \psi_2}$ **return** $par(R_{I_1}, Sat(M, \psi_1)) \cup par(R_{I_2}, Sat(M, \psi_2))$

else if $\phi \equiv \mathbf{EX}(\psi)$ **return** $\{s : (s, s') \in R, s' \in Sat(M, \psi)\}$

else if $\phi \equiv \mathbf{EG}(\psi)$ **return** $vZ.(Sat(M, \psi) \cap \mathbf{EX}(Z))$

else if $\phi \equiv \mathbf{E}(\psi_1 \mathbf{U} \psi_2)$ **return** $\mu Z.(\boxed{par(R_{I_1}, Sat(M, \psi_1)) \cup par(R_{I_2}, Sat(M, \psi_2))} \cap \mathbf{EX}(Z))$

end if

By now, we consider the access to the atomic propositions transparent to the underlying technology. However, we must advance that this is usually the most time-consuming part during the experimentations in Section 4. This fact motivates the introduction of information systems optimized for the management of huge amounts of phylogenetic data in the next section.

Finally, the size of the slices depends on the target DNA regions we desire to analyze (i.e., single nucleotides, genes or chromosomes), the kind of properties we want to verify and the hardware requirements we have. A high number of slices will provide of a better level of parallelism and low hardware requirements (CPU, memory), but it will be limited by the potential appearance of bottlenecks during the composition of results. The detection of backmutations is a degenerated example of this case because it only needs a tree labeled with a single nucleotide.

3   Database Model Checking

The second step of our approach would consist of uncoupling the atomic propositions from the model checker. The use of external databases as a repository of nucleotide labels alleviates the memory explosion problem when the local storage of trees with partial DNA is not enough. Moreover, the database manager simplifies the interface to the DNA data because it usually allows concurrent queries and it hides the internal synchronization and data structures. In a general sense, trees are labeled with pointers to DNA sequences stored in an external server.

By default, we use BioSQL, a shared database schema for storing sequence data in SQL servers that is supported by several script languages [9]. The DNA alignment is stored in a single table with a row per DNA sequence. Each row consists of two fields: an identifier (GenBank accession number) plus the plain string of nucleotides.

In fact, the database tables can be seen as matrices that we can slice by row (number of taxons) or by column (divide the DNA in substrings). These (sub)tables can be stored or replicated in separated servers or cluster nodes so as to allow parallel access to the DNA data and to improve the communication bandwidth between the database and the model checker. In addition, recent versions of SQL servers support multicore CPU's, which improves the time response when attacking the server with several queries. This approach allows scaling in memory and speeds up the system.

Finally, relational databases cannot only store phylogenetic data, but also partial verification results. Furthermore, relational databases can execute the model checking algorithms for CTL formulas [12]. The main advantage is that we avoid the exportation of DNA data from the database and the bandwidth bottleneck. An evaluation of SQL model checkers for phylogenetic data will comprise our future work.

4   Experimental Results of the Workflow

We compose our new framework by integrating the different modules described in previous sections (Figure 1). The central core of our approach is the model checker package NuSMV v2.5.4 [7], which is a well-known public software. It is surrounded by a set of modules and tools that pre (post) process the phylogenetic data. Our framework is divided in three important modules. First of all, the "Loader" consists of a Biopython script that creates and initializes a local MySQL server v5.5.24-7 with all the DNA sequences. It is executed only once during the initialization and the database remains constant for all the verification process.

Secondly, the "Property Transformer" is a Biopython script that translates the phylogenetic formulas and precomputes the set of states (usually, species) satisfying the atomic propositions (it invokes the last recursion call of Algorithm 1). We have selected to precompute the set of atomic propositions in parallel because they are the most time-consuming part of the verification process. The script takes an atomic proposition with the pattern $seq[i] = \sigma$ and returns the set of identifiers from those species that satisfies the formula. We recall that the internal representation of the phylogenetic tree in the model checking tool will be labeled with unique GenBank identifiers that allow us to recover the DNA string from the database. Finally, the "Property Transformer" generates a property file per atomic proposition so as to launch several copies of the model checker in parallel.

Next, the "NuSMV Transformer" is a Bioperl script that translates the phylogenetic tree into a NuSMV file format and appends the properties we want to verify. The NuSMV input file is equivalent to that defined for Cadence SMV in [10] except for the inclusion DNA sequences. The script can be extended in order to divide the original tree in multiples subtrees according to the biological requirements. Finally, for each (sub)tree, the system executes in parallel the verification of the atomic propositions and composes the result with a multi-threading algorithm.

With respect to the experimentation, we must remark that the integration of all the techniques explained in previous sections (external databases and distribution) enables the verification of properties in big phylogenies. Our data set comprises the ZARAMIT tree [4], which has been reconstructed from 7390 Human mitochondrial DNA sequences (16,569 base pairs). Table 1 shows that our architecture outperforms previous works in [10]: we spend less time and memory for the initialization of the model checking tool with the ZARAMIT tree than for the initialization of a protein tree (2000 nodes and sequences of 500 aminoacids).

We must emphasize that we cannot execute the model checking tool if we do not separate the DNA data in an external database. From the point of view of performance, the database is the most important point of our workflow because it connects the model

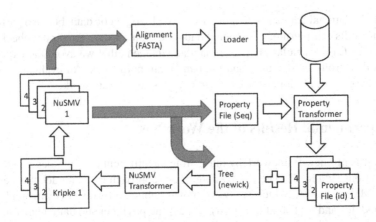

Fig. 1. Workflow diagram

checker with the DNA sequences. The database server is executed in a desktop workstation (AMD Opteron @3GHz, 4GB RAM, Debian Linux). Due to the "small" alignment size (around 350MB for 7390 leafs plus ancestors) and the hardware available, we work with a single instance of MySQL server. The database loader spends less than 2 minutes at the initialization point, but the database remains constant for all the verifications. For example, the retrieval of atomic propositions lasts 7.2s for 25 concurrent SQL queries (44.6s if we serialize them).

A second desktop workstation (Intel Core 2 Duo E6750 @ 2.66GHz, 8GB RAM, Debian Linux) is devoted to the execution of the model checking workflow. As we remarked previously, we detected that the internal representation of the Kripke structure in NuSMV penalizes the access to the atomic propositions during the model checking process. The time required for the computation of a relatively big set represented by a single atomic proposition raises up to 1m 20s (around 25s if we discount the initialization costs of Table 1). Conversely, the integration of partial results (atomic proposition sets) and the computation of CTL paths for the verification of a backmutation property only lasts 1m 2s in total (5-10s if we exclude the initialization costs).

The next step consists of the distribution of the computation of sets satisfying the atomic propositions. We launch an independent NuSMV task per atomic proposition so as to obtain as much parallelism as possible. The first part of the parallelism consists of the replication of the Kripke structure in several instances, the verification of each atomic proposition in a monolithic model checker, and the storage of the temporal result (sets satisfying an atomic proposition) in a text file. Thanks to the integration with databases, the memory consumption of each model checker is reduced (a small overhead of around 50 MB/task for the representation of the phylogenetic tree). Besides, they can be executed in multicore CPU's or clusters.

When every task is finished, we execute a parallel version of the model checker so as to collect the results of the atomic propositions and analyze the CTL paths. Depending on the number of formulas and the number of atomic propositions per formula, we obtain a higher speed up from the distribution. We remind that the computation of sets represented by atomic propositions lasts around 25s in usual cases, and the initialization costs raises up to 55s-1m, so we must parallel the computation of four or more atomic propositions in order to obtain a clear speed up. Longer unexplored phylogenetic properties can take advantage of multi-threading as the CTL verification cost depends of the formula length [1].

5   Conclusions

The use of monolithic model checking for the verification of huge phylogenetic data is unfeasible because of the high requirements in time and memory. In this paper we presented the adaptation and integration of two techniques that help us to scale up our framework: distributed model checking, that divides the Kripke structure in subtrees and slices the nodes in portions of DNA; and the use of external databases.

The particularities of trees with respect to more generic graph structures facilitates the implementation, synchronization and composition of results of distributed structures in phylogenetics. The verification of properties in clades and subtrees improved the performance of our system as the initialization time in the NuSMV model checker depends quadratically with the number of nodes.

Secondly, the introduction of sliced model checking has reduced the memory consumption by storing fragments of the DNA in the local memory of the model checker or the complete DNA alignment in an external persistent database. In particular, relational databases allowed us to map sequences and nucleotides with atomic propositions. Furthermore, relational databases can emulate the model checking algorithms using PL-SQL. In conjunction with a multi-threading parallel algorithm, we have provided an efficient way of verifying complex temporal logic formulas.

Finally, the integration of all this approaches helped us to execute the verification of properties in a real phylogenetic tree. The execution of model checking over the ZARAMIT tree has outperformed the results of monolithic model checking in [10]. Our future work will focus on the symbolic manipulation of results and the evaluation of a set of phylogenies labeled with quantitative information so as to reason which tree fits a DNA substitution model expressed in a probabilistic CTL.

Acknowledgments. This work was supported by the Spanish Ministry of Science and Innovation (MICINN) [Project TIN2011-27479-C04-01] and the Government of Aragon [B117/10].

References

1. Baier, C., Katoen, J.-P.: Principles of model checking. The MIT Press, Cambridge (2008)
2. Benson, D.A., Karsch-Mizrachi, I., Clark, K., Lipman, D.J., Ostell, J., Sayers, E.W.: GenBank. Nucleic Acids Reseach 40, D48–D53 (2012)
3. Blanco, R., de Miguel Casado, G., Requeno, J.I., Colom, J.M.: Temporal logics for phylogenetic analysis via model checking. In: 2010 IEEE Int. Conf. on Bioinformatics and Biomedicine Workshops, pp. 152–157. IEEE (2010)
4. Blanco, R., Mayordomo, E., Montoya, J., Ruiz-Pesini, E.: Rebooting the human mitochondrial phylogeny: an automated and scalable methodology with expert knowledge. BMC Bioinformatics 12, 174 (2011)

5. Bošnački, D., Edelkamp, S.: Model checking software: on some new waves and some ever-greens. Int. J. Software Tool Tech. Tran. 12, 89–95 (2010)
6. Boukala, M.C., Petrucci, L.: Distributed CTL Model-Checking and counterexample search. In: 3rd Int. Workshop on Verification and Evaluation of Computer and Communication Systems (2009)
7. Cimatti, A., Clarke, E., Giunchiglia, E., Giunchiglia, F., Pistore, M., Roveri, M., Sebastiani, R., Tacchella, A.: NuSMV 2: An OpenSource Tool for Symbolic Model Checking. In: Brinksma, E., Larsen, K.G. (eds.) CAV 2002. LNCS, vol. 2404, pp. 359–364. Springer, Heidelberg (2002)
8. Fitch, W.M.: Uses for Evolutionary Trees. Philosophical Transactions of the Royal Society of London. Series B: Biological Sciences 349, 93–102 (1995)
9. Mangalam, H.: The Bio* toolkits a brief overview. Brief Bioinform. 3(3), 296–302 (2002)
10. Requeno, J.I., Blanco, R., de Miguel Casado, G., Colom, J.M.: Phylogenetic Analysis Using an SMV Tool. In: Rocha, M.P., Corchado Rodríguez, J.M., Fdez-Riverola, F., Valencia, A. (eds.) PACBB 2011. AISC, vol. 93, pp. 167–174. Springer, Heidelberg (2011)
11. Requeno, J.I., Blanco, R., de Miguel Casado, G., Colom, J.M.: Sliced Model Checking for Phylogenetic Analysis. In: Rocha, M.P., Luscombe, N., Fdez-Riverola, F., Corchado Rodríguez, J.M. (eds.) 6th International Conference on PACBB. AISC, vol. 154, pp. 95–104. Springer, Heidelberg (2012)
12. Shegalov, G.: CTL Model Checking in Database Cloud. Unpublished version paper (2011)

Comp2ROC

R Package to Compare Two ROC Curves

Hugo Frade[1] and A.C. Braga[2]

Engineering School, University of Minho, Campus Gualtar, Braga, Portugal
[1] MSc. Student in Informatics Department
hugofrade@me.com
[2] Production and Systems Engineering Department
acb@dps.uminho.pt

Abstract. This paper describes the Comp2ROC package implemented in the R programming language. The theoretical contextualization behind this package are introduced including a general introduction to ROC curves, their main features, and how ROC curves are applied. Furthermore, methodologies are presented for comparing two ROC curves, both when two curves intersect and when they do not. Finally, the paper explains how and when to use the Comp2ROC package to compare two ROC curves that intersect. An example application is shown and discussed to fully demonstrate how to use the facilities in this package.

Keywords: ROC Curve, multi-objective, R Package,Comp2ROC.

1 Introduction

The ROC (Receiver Operating Characteristic) curves are used to evaluate the accuracy of a prediction. These curves were first used in Second World War, on the analysis of radar signals. After the attack on Pearl Harbor, the Army of the United States began its research to improve the prediction of correctly detect Japanese planes through these radar signals [7].

Systems comparison based on ROC curves is a process widely used the most varied areas of knowledge. In two ROC curves that do not intersect the statistic proposed by Hanley [3] is sufficient to evaluate the statistical significance. However when two ROC curves intersect is needed to make a comparison according to these crosses as well as identify the region where they occur. The methodology proposed by Braga et al [1] lets the user compare two ROC curves that intersect through a nonparametric method. The Comp2ROC is a package developed in R language [6], which is based on this new methodology.

The choice of the R language relates to the fact that this is a language with an integrated development environment for statistical calculations and graphics.

M.S. Mohamad et al. (Eds.): *7th International Conference on PACBB*, AISC 222, pp. 127–135.
DOI: 10.1007/978-3-319-00578-2_17 © Springer International Publishing Switzerland 2013

2   Implementation

2.1   Method Description

ROC curve is a representation of a set of points in an unitary two-dimensional space, which are joined by straight lines. The ROC graph ordinate axis is *Sensitivity* or TPR and the abscissa axis is $1 - Specificity$ or FPR, obtained by varying the cut-off along the axis of the decision variable in a discrimination problem.

The important parameters for accuracy evaluation are the predictions results and the observed results (binary values). This decision process could be described as a function of the observed and predicted values, as a table 2x2 table as following:

Table 1. Relation between previewed and seen data.

Predicted	Observed Positve	Negative
Positive	TP (True Positives)	FP (False Positives) Type I error
Negative	FN (False Negatives) Type II error	TN (True Negatives)

The values above allow calculating the sensitivity and specificity. The sensitivity (TPR) measures the ability of a test to correctly classify an individual as having a specific attribute and is calculated as follows:

$$Sensitivity = \frac{TP}{TP + FN}. \tag{1}$$

On the other hand, the specificity (TNR) measures the ability of a test correctly to exclude an individual who does not have a particular attribute, and is calculated as follows:

$$Specificity = \frac{TN}{TN + FP}. \tag{2}$$

Discrimination Measures of the Curve. As discrimination measure of the curve, are used the area below it (A). It value can be computed using:

1. The approximation to the Wilcoxon-Mann-Whitney statistic:

$$S(X_i, Y_j) = \begin{cases} 1 & \text{if } X_i > Y_j \\ 0.5 & \text{if } X_i = Y_j \\ 0 & \text{if } X_i < Y_j \end{cases} \tag{3}$$

where X_i are the negative cases, Y_j are the positive cases, and S is the classifier function. The area, A could be compute using the equation:

$$A = \frac{1}{n_N \times n_P} \sum_{i=1}^{n_N} \sum_{j=1}^{n_P} S(X_i, Y_j) \qquad (4)$$

2. Numerical algorithms as trapezoidal rule;
3. Approximation formula in a binormal plan, developed by Metz [4].

Compare ROC Curves That do not Intersect. In Fig. 1 are illustrated three ROC curves that do not intersect and the randomness line (dashed line). A ROC curve reaches its gold standard when it overlaps with the upper left corner of the referential, represented in Fig. 1.

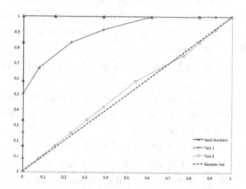

Fig. 1. Example of three ROC curves.

Two curves could be compared seeing which one has the largest value of A. The biggest area corresponds to the best performance. To evaluate the statistical significance of the difference between two curves the methodology proposed by Hanley [3] through the formula (5), is the most used.

$$Z = \frac{A_2 - A_1}{\sqrt{SE_1^2 + SE_2^2 - 2rSE_1SE_2}} \sim N(0,1) \qquad (5)$$

A_i is the area below the curve i, SE_i is the standard error of the area of curve i and r is the correlation coefficient between the two curves and is calculated through the method developed by DeLong [2]. If the samples are independent $r = 0$.

Compare ROC Curves that Intersect. The comparison of ROC curves that cross each other as proposed by Braga et al [1] is done in three main phases:

1. **Sampling curves.** Are drawn K sampling lines that vary their slope uniformly covering the entire referential. These lines start on point $(1,0)$ and

progress in the direction of the ordinate axis. The authors found an optimal number of sampling lines, but these are computationally impossible to process, so they propose 100 sampling lines as the optimal number.

2. **Curve performance evaluation**. To evaluate the measures of performance of the curves, are computed two measures, extension and location. **Extension measure** allows to evaluate the proportion of gains and losses. Is checked line and curves intersection points, and the first curve that intersects the line gains. **Location measure** is related to the area bounded by the two curves and two sampling lines. The area is calculated through the difference between the triangle areas formed by the two intersection points and origin point (1,0). After is drawn a chart with these values. All positives correspond to curve 1 gain, while the negatives correspond to a curve 2 gain. The region with highest area expressivity corresponds to the best performance.

3. **Statistical evaluation of the performance of the modalities to compare.** In this phase a permutation test is carried out. To do this permutation test Braga et al [1] proposed:
   (a) Formulate the null and alternative hypothesis, and choose the test statistic, TS;
   (b) Calculate the TS for the original set of observations;
   (c) Permute the observations and we calculate the TS for all combination;
   (d) Obtain the critical point of the TS;
   (e) Decision to reject or not the null hypothesis.

   In this case it was tested if the differences are or not significant, using as test statistics (TS) the sum of the differences defined by the authors [1].

2.2   Package Description

Version 1.0 is the result of compiling functions that implement a multi-objective optimization methodology for evaluating two diagnostic systems based on ROC curves performance. The package allows drawing ROC curves in the unitary ROC plan, distances graphs and see the curves comparison results. The application is developed in R (requires version 2.15.1 or higher and **boot** and **ROCR** packages). Next will be described the package functions used on Comp2ROC.

Curvesegslope Function. This function calculates the slopes of the lines that together build the ROC curve. It takes as parameters the TPR and FPR of the ROC curve.

Curvesegsloperef Function. Calculates the slopes of the lines that join the points of the ROC curve to the reference point (1,0). The function parameters are the TPR, the FPR and also the reference point.

Lineslope Function. Calculates the slopes of the sampling lines, depending on the number desired by the user (K). It returns a vector with these values.

Linedistance Function. This function calculates the intersection points of the sampling lines with the ROC curve. It also calculates the distance between these points and the reference point. The parameters of this function are the TPR and FPR points, the slopes calculated by *curvesegslope, curvesegsloperef, lineslope* and also the reference point.

Areatriangles Function. This function calculates the triangles areas formed by two sequential points and the reference point. In addition it also calculates the total area, based on previous triangles. The parameters of this function are the results obtained in *lineslope* and *linedistance*.

Diffareatriangles Function. This function calculates the difference between triangles areas of the two curves. This also allows calculating the difference between the total areas. The parameters of this function are the triangles areas of the two curves obtained in *areatriangles*.

Rocsampling Function. Creates the two curves and compare them by extension and location. All functions above are used here. It takes the TPR and FPR points of each ROC curve and also the number of sampling lines, K. As output value, returns a list of the overall areas, the proportions and the locations for each curve, the slopes of the sampling lines, the differences between areas and finally the distances calculated by *linedistance*.

Rocsampling.Summary Function. Receives and presents the results obtained through the function *rocsampling*. These data are presented in the R command line using the structure presented in Fig. 2(a).

Comp.Roc.Curves Function. This function calculates by bootstrapping the real distribution for the entire set. It takes as parameters the result obtained by *rocsampling*, two flags and the name to the plot. One flag indicates if the user wishes calculate the confidence interval and the other if the user wants to make the plot. As output values, it returns the test statistic, two p-values (one-sided and two-sided) and the confidence interval.

Comp.Roc.Delong Function. This function is related to the areas calculation and some statistical measures. Firstly, divides the data into two categories, negative and positive. Then calculate the Wilcoxon Mann Whitney matrix for each modality. Next are calculated some values such as areas, standard errors and global correlations, which are the output of the function. It takes as parameters the data of each curve and their status.

Roc.Curves.Plot Function. This function draws the graph of the two ROC curves. Curve 1 is shown by solid line and curve 2 is shown by dotted line. The random line is shown by dashed line. This graph was illustrated in Fig. 2(b).

Read.Manually.Introduced Function. This function reads the data that will be used in comparison. Only works with data already introduce in the R workspace. It needs the name of two modalities and their dependency. If they are independent the user must give second modality status. This variable must be ordered by state, first negatives values followed by the positives. At the end the user must identify the test direction for each modality.

Read.File Function. Reads the data that will be used in comparison, like *read. manually.introduced*, but in this case only works with data saved in `txt` and `csv` formats, so the user must say if there's any header in the file. In addition to the parameters already described in *read.manually.introduced* is also needed to indicate the column and decimal separator.

Roc.Curves.Boot Function. This is the core function of the package. The user must give the data, an α for the confidence level, the graphics display name and the number of permutations. Apart this, is also needed the name of the two modalities that will be compared. As return value, it give a list with the areas of each modality, standard errors, confidence intervals, the areas through the trapezoidal rule, the correlation coefficient, the sum of the differences of areas and corresponding confidence interval, the Z-stats and its p-value and the number of existing crossings.

Rocboot.Summary Function. This function displays the data obtained in *roc. curves.boot* in the R command line. These are structured according to Fig. 2(e).

Save.File.Summary Function. This function saves the results as the structure shown in *rocboot.summary function* in a `txt` file. To save this data the users must provide the results obtained, the file name and also the names of the two modalities as well a parameter that indicates the way that the file is saved (overwrite or attach).

3   Results and Discussion

In this section is explained how to use Comp2ROC and some final results examples. We test the package using data created by Zhang [8].

The comparison test should be done using the following commands in the order they are presented:

1. `nameE="new_Zhang"` - graphs and output file name
2. `moda1="mod1"` - modality one column name
3. `moda2="mod2"` - modality two column name

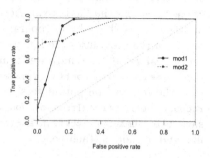

(a) First results obtained

(b) ROC curves and random line Plot

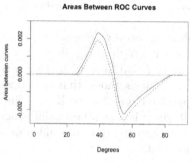

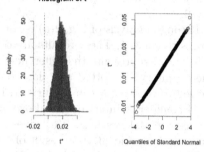

(c) Area Plot

(d) Distribution Plot. In the left chart we have a vertical dashed line that represents the position of the distribution test statistic (TS).

```
---------------------------------------------
mod1
---------------------------------------------
Area:                                0.919775
Standard Error:                      0.006675
Area through Trapezoidal Method:     0.919710
CI Upper bound (Percentil Method):   0.933964
CI Lower bound (Percentil Method):   0.907014
---------------------------------------------


---------------------------------------------
mod2
---------------------------------------------
Area:                                0.925818
Standard Error:                      0.004649
Area through Trapezoidal Method:     0.925756
CI Upper bound (Percentil Method):   0.934286
CI Lower bound (Percentil Method):   0.917018
---------------------------------------------

Correlation Coefficient between areas: 0.604370

TEST OF DIFFERENCES
Z stats:  -1.128807
p-value:   0.258979

Sum of Global Areas Differences:  -0.006043
CI Upper bound (Percentil Method):  0.009335
CI Lower bound (Percentil Method): -0.021778
Number of Crossings: 1
```

(e) Final Results

Fig. 2. Comp2ROC output images

4. `data=read.file("exemplo_zhang.csv",TRUE,";",",",moda1,TRUE,moda2,TRUE,"status",TRUE)` - the file is read. According to the details, this file has header and the ";" is the column separator and "," is the decimal separator. The names of the modalities are followed by the direction of the test. At the end the logical value True, indicates that the modalities are related.
5. `results=roc.curves.boot(data,name=nameE,mod1=moda1,mod2=moda2)` - calculates the comparison parameters
6. `rocboot.summary(results,mod1,mod2)` - the results are shown in command line
7. `save.file.summary(results,nameE,app=TRUE,mod1,mod2)` - results are saved in file, and in mode attachment

If the user wishes to use data that have been entered manually instead of the command `read.file` the user must use the command `read.manually.introduced`.

The R results are presented while the steps are performed, as illustrated in Fig. 2.

Through analysis of Fig. 2(a) is concluded that, in terms of proportion, curve 2 wins to curve 1. These results are confirmed in the graph in Fig. 2(b).

Fig. 2(c) represents the difference between areas (solid line) with their confidence intervals (dotted and dashed line). The analysis shows that all values above 0 represents a win to curve 1, in the opposite curve 2 wins. If both limits of confidence intervals are above or below 0, there is a significant difference between modalities. Fig. 2(d) gives $t*$ (test statistic) distribution and also a Quantile-Quantile chart as result of permutation test.

Final results are shown in Fig. 2(e). This shows the results for each modality to be compared individually, in terms of area values under the ROC curve, their standard errors, and confidence intervals for each corresponding area. These results confirm that in terms of area under ROC curve the, the modality 1 performs better than 2. The value of correlation coefficient between areas reveals that two modality are related. In the comparison tests results, the value of Z-stats and respective p-value, indicate that on the whole there are no statistically significant differences between the two modalities. The values obtained for the two limits on confidence interval for the difference between areas (proposed method) also indicate that on the whole there are no statistically significant differences between the two modalities. However the information in the graph of Fig. 2(c) allows detect two distinct regions were one curve performs better than other.

4   Conclusions and Future Work

This package answer the problem of comparing two ROC curves that cross each other according the methodology proposed by Braga et al [1]. This package also allows the user to visualize graphically the region on ROC space where one curve is superior to another. All code and associated functions have resulted in one package which is designated Comp2ROC. The notes and examples were created from an example in the literature [8].

There is some future work that will be implemented in further versions. At this point, the user just needs to use seven steps to perform the functionality of this. One of the changes to do is divide the function *roc.curves.boot* into smaller functions for intermediate values could be saved. We'll also try to display the intersection points to allow the user to identify the coordinates of these points in ROC space.

References

1. Braga, A., Costa, L., Oliveira, P.: An alternative method for global and partial comparasion of two diagnostic system based on ROC curves. Journal of Statistical Computation and Simulation (2011)
2. DeLong, E., DeLong, D., Clarkepearson, D.: Comparing the areas under 2 or more correlated receiver operating characteristic curves - a non parametrical approach. Biometrics 44(3), 837–845 (1988)
3. Hanley, J., McNeil, B.: A method of comparing the areas under receiver operating characteristic curves derived from the same cases. Radiology 148(3), 839–843 (1983)
4. Metz, C.E.: Statistical Analysis of ROC Dara in Evaluating Diagnostic Performance. In: Proceeding of First Midyear Topical Symposium: "Multiple Regression Analysis Application in the Health Science", vol. 13, pp. 365–384 (1986)
5. Pepe, M.S.: The Statistical Evaluation of Medical Tests for Classification an Prediction. Oxford Statistical Science Series. Oxford University Press, New York (2003)
6. R Development Core Team. R: A language and environment for statistical computing. Vienna: R Foundation for Statistical Computing (2006), http://www.R-project.org
7. Swets, J.A., Pickett, R.M.: Evaluation of Diagnostic Systems Methods from Signal Detection Theory. Academic Press, London (1982)
8. Zhang, D., Zhou, X., Freeman, D., Freeman, J.: A nonparametric method for the comparison of partial areas under ROC curves and its application to large health care data sets. Stat. Med. 21(5), 701–715 (2002)

Network Visualization Tools to Enhance Metabolic Engineering Platforms

Alberto Noronha[1], Paulo Vilaça[2], and Miguel Rocha[1]

[1] CCTC, School of Engineering, University of Minho, Portugal
mrocha@di.uminho.pt
[2] SilicoLife, Lda, Taipas, Portugal
pvilaca@silicolife.com

Abstract. In this work, we present a software platform for the visualization of metabolic models, which is implemented as a plug-in for the open-source metabolic engineering (ME) platform OptFlux. The tools provided by this plug-in allow the visualization of the models (or parts of the models) combined with the results from operations applied over these models, mainly regarding phenotype simulation, strain optimization and pathway analysis. The tool provides a generic input/ output framework that can import/ export layouts from different formats used by other tools, namely XGMML and SBML. Thus, this work provides a bridge between network visualization and ME.

Keywords: Metabolic models, biological network visualization, metabolic engineering, open-source software.

1 Introduction

Bioinformatics has recently become an important discipline in life sciences research, mainly due to the high amount of experimental data generated by high throughput sequencing and omics approaches. These data are stored in different databases, in multiple formats, bringing problems in their integration. Due to the difficulty of interpreting and analysing large-scale databases, networks became useful tools for understanding the complexity of biological processes, being commonly used to represent biological entities and their interactions. Indeed, networks are used to represent metabolic pathways, regulatory information and signal transduction systems[3].

Over the years, several tools arose to address the complexity of analysis and visualization tasks for large-scale networks. The use of graphs is very common, due to the easy representation of biological entities as nodes and their interactions as edges. The fact that each entity may be connected with several others highlights the importance of the multi-edged networks approach that several tools offer [10]. The main goal of any visualization tool is the capability of identifying or highlighting patterns and other information not easily available through the analysis of the raw data.

The ability to visually represent metabolic networks is already provided by several tools whose enumeration is out of the scope of this paper. In this regard, we would

M.S. Mohamad et al. (Eds.): *7th International Conference on PACBB*, AISC 222, pp. 137–144.
DOI: 10.1007/978-3-319-00578-2_18 © Springer International Publishing Switzerland 2013

emphasize *Cytoscape* [14] that has emerged as a powerful and extendible tool for network analysis and visualization. While these tools already exist, they are not currently connected with any of the few existing Metabolic Engineering (ME) frameworks.

Some related efforts in the modelling of biological systems have been key tasks in the computational systems biology field, whose aim is to develop and use algorithms, data structures and tools to allow the integration of data into models that can be used to simulate the behaviour of biological systems. In particular, the development of metabolic models and tools to simulate the behaviour of these systems, allows for the creation of important tools in the context of ME, whose main goal is to optimize cellular metabolism deriving rational genetic modifications that lead to strains capable of producing compounds of interest.

While for the dynamic modelling of large-scale systems, it is often necessary to have mechanistic detail and kinetic parameters not commonly available, structured-oriented analyses only requires the usually well-characterized network topology [15]. Constraint-based analysis methods that consider network stoichiometry, and potentially other constraints such as maximal pathway capacities, are proving to be very reliable ways to obtain a quantitative perspective of various aspects of cellular function. In particular, these have gained considerable popularity for simulating cellular metabolism, using methods such as flux balance analysis (FBA) [11] to provide phenotype simulation and as the basis for strain optimization [13].

The use of these methods in the context of ME has provided a number of valuable algorithms and tools. The authors research group has been active in this line of research and the main results are available for the ME community in the form of an open-source software platform, *OptFlux* (http://www.optflux.org) [12]. This application includes a number of tools for ME, spanning phenotype simulation methods for wild type and mutant strains, strain optimization algorithms, support for several model formats, pathway and topological analysis, among others.

In this context, the main aim of this work is to develop computational tools for the visualization of metabolic networks, in the context of a metabolic engineering platform (*OptFlux*), providing an adequate integration between the features of both tools. The overall idea is to provide visualization capabilities to *OptFlux*, powering this tool with extended visual analysis capabilities. In more detail, the following technological objectives will be pursued:

- Implement a generic library for the visualization of metabolic models or parts of these models;
- Implement tools to allow importing/exporting model layouts in standard formats, such as *SBML* [8] and *XGMML* [2];
- Integrate those tools in a ME platform, integrating data obtained from phenotype simulations, strain optimization and pathway analysis, and allowing the exportation of the results in the standard formats.

2   Implementation

Metabolic models collect relevant metabolic information in a mathematical format. ME makes use of these models to predict ways to maximize the production of desired compounds, through directed genetic changes. Usually, this task is done through phenotype simulations and maximizing desired fluxes from the reactions in the model. The visual representation of these networks should be as simple as possible, to facilitate naked-eye interpretation. Considering the nature of ME simulations, which generate flux values distributions, for given conditions, the network can then be viewed as a set of reactions, and a simulation results as a set of values for those reactions.

Taking this into account, a visual representation of a metabolic model was envisioned. As expected, the network is represented as a graph, where biological entities (reactions and metabolites) are represented as nodes, and their interactions (consumption and production) as edges. The architecture of the visualizer library has two distinct layers (Figure 1): the *BiovisualizerCore* contains the specification of a visual representation of the metabolic network (layout), the tools to represent it in a 2D interface and methods for visual manipulation of the network; the second layer, the Input/Output layer, consists in the readers and writers for a variety of formats that can provide layouts to the *BiovisualizerCore*, working as a transformer receiving a file in a specific format, reading it and generating a layout.

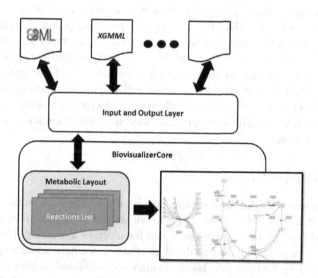

Fig. 1. Visualizer architecture: I/O layer imports/ exports standard network files and transforms them into metabolic layouts; the *BiovisualizerCore* layer has the capability to visually represent the metabolic layouts.

2.1 BioVisualizerCore

The *BiovisualizerCore*'s main purpose is to transform a metabolic layout into a 2D graph. The metabolic layout definition is represented by a list of reactions and was implemented as a Java interface. Each reaction has its identifier, spatial coordinates (x,y), reversibility information, additional information, a list of reactants and a list of products. Each metabolite has information on its identifier and coordinates. Therefore, there are three types of nodes: reaction, metabolite and information nodes. The visual representation of the network is handled by the *Prefuse* library, a graph visualization framework written in Java [1].

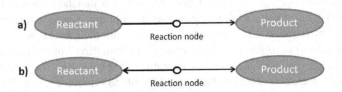

Fig. 2. *BiovisualizerCore*s representation of metabolic reactions. Each reaction has a reaction node, and reactants and products: a) irreversible reaction; b) reversible reaction.

Each node of the graph can have its 2D coordinates defined, but they are not mandatory. For the nodes that do not have a position, an automatic layout provided by *Prefuse* is applied, called *Force Directed Layout*, that positions graph elements based on a physics simulation of interacting forces; by default, nodes repel each other, edges act as springs, and drag forces are applied. *BiovisualizerCore* packages provide a series of functionalities to enhance the interpretation of the network. Hovering the cursor over a node highlights it; if the node is a reaction, it will also highlight its edges and neighbour nodes. It has a series of visual filters, allowing to change the visibility of nodes/ edges based on attributes. Finally, it also provides ways to change the thickness of edges based on numerical values.

2.2   Input/Output Layer

BiovisualizerCore gives the possibility to visually represent a metabolic layout, but building that layout is still needed. There are several tools that allow the creation of networks and exportation of those networks for several file formats. The I/O Layer has the objective of providing the capability to read a given network in a specific file format, and building the metabolic layout, readable by *BiovisualizerCore* packages. At the same time it shall also provide the possibility to export those metabolic layouts into those specific formats.

Currently, this layer supports two formats: *CellDesigner SBML* and *XGMML*. *CellDesigner* [6] is a well-known tool for drawing cellular networks, based on the process diagram paradigm, with a graphical notation system proposed by Kitano. These are stored using an extension of the *Systems Biology Markup Language (SBML)*. On the other hand, *XGMML (eXtensible Graph Markup and Modeling Lan-guage)* [2] is an extension to *GML (Graph Markup Language)*, and it is used for graph description using XML tags to describe nodes and edges of a graph, being supported for instance by *Cytoscape*.

2.3   OptFlux Interaction

BioVisualizerCore packages are integrated within *Optflux* to provide new plug-ins for this framework. In terms of its implementation, *Optflux* is built on top of *AIBench* [7], a framework for the development of scientific applications, bringing important advantages to both the developers and the users, given its design principles and archi-tecture. The applications follow the MVC (model-view-controller) pattern and they are plug-in based facilitating the reuse and integration with additional developments. We designed an intuitive connection between layouts and metabolic models, map-ping identifiers of the nodes with identifiers of the respective compounds/reactions in the model. Note that the layouts can contain only a part of the entities in the model, being also possible to create distinct layouts for a given model.

Using this connection it will be possible to access metabolic information directly from the visualizer by clicking the nodes. Visual filters include hiding information nodes, hiding fluxes not present in the metabolic model and hiding reactions with zero value fluxes, given a flux distribution. Flux distributions can be obtained from phenotype simulations performed in *OptFlux*, bringing the capability to visualize the effect that those simulations have on the network, by changing the thickness of the edges of the reactions according to that simulation, by normalizing the value of the fluxes. The *I/O Layer* supports distinct file formats, and to integrate it with *OptFlux* and facilitate the addition of new formats, we decided to implement a series of *OptFlux* plugins. Currently, there are two plugins for the visualization of metabolic networks:

- *CellDesigner plugin*: supports *CellDesigner SBML* created layouts. The SBML reader used is the same *OptFlux* uses to read SBML models using JSBML, a Java library for reading, writing, and manipulating SBML files and data streams [4].
- *XGMML plugin*: gives support for networks created in the XGMML format. *Cytoscape* [14] is one of the tools that support the creation of these networks. The exportation works in two different ways: in the first, the layout imported is preserved, while in the other one exports the *BiovisualizerCore* layout appearance.

3   Case Study

To illustrate some of the features of the tool, we loaded a simplified model for Saccharomyces cerevisiae [5]. A *CellDesigner* SBML layout was also loaded.

Figure 3a) shows the overview of the network, with all nodes visible; 3b) shows the highlighted biomass reaction, with all its reactants and products also highlighted; 3c) shows metabolite CO_2 highlighted, and since this metabolite appears in several reactions, those nodes are also highlighted; finally, 3d) shows a flux distribution obtained from simulating a wild type strain. The thickness of the edges is changed, the filter to hide zero value fluxes is activated and the labels of the reactions are changed by adding the value of the flux after the reaction name.

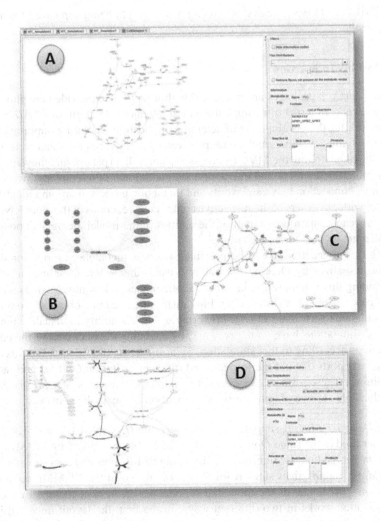

Fig. 3. Some functionalities of the visualizator. A. network overview; B. biomass reaction highlighted; C metabolite CO2 highlighted; D Flux distribution.

4   Conclusions and Future Work

BiovisualizerCore packages and *Optflux* plugins provide a freely available tool that allows users to visualize the results of *in silico* phenotype simulations, which is something new in ME tools. The ability to export those results and import them in other tools (e.g. *Cytoscape*) is also a valuable asset. One of the major applications of these features is the enhanced ability to analyse results of strain optimization results. In this scenario, these tools allow looking at simulations of the best mutants found by the strain optimization process and analysing their phenotype, thus visually searching for clues that can highlight the strategies followed to produce a given compound.

Also, these tools are linked with the elementary modes analysis plugin (EFM4Opt-Flux) providing a visualization tool for the set of elementary modes found for a given network, which can be then be exported for instance to *CellDesigner* or *Cytoscape*.

There are some aspects of the visualizer that can be improved in the future. Allowing the users a broader configuration of the aspect of the network is something worth exploring. Improving the I/O layer by providing a larger spectra of format support is another, for instance integrating KEGG [9] layouts through the format KEGG-ML.

OptFlux is freely downloadable from the website http://www.optflux.org and both plugins cited in this work are available through the plugin manager the software provides in the *Help* menu.

Acknowledgements. The work is partially funded by ERDF - European Regional Development Fund through the COMPETE Programme (operational programme for competitiveness) and by National Funds through the FCT (Portuguese Foundation for Science and Technology) within projects ref. COMPETE FCOMP-01-0124-FEDER-015079 and PEst-OE/EEI/UI0752/2011.

References

1. Prefuse libraries (2013), https://github.com/prefuse/Prefuse
2. Xgmml specification (2013), http://cgi5.cs.rpi.edu/research/groups/pb/punin/public_html/XGMML/
3. Barabási, A.-L., Oltvai, Z.N.: Network biology: understanding the cell's functional organization. Nature Reviews. Genetics 5(2), 101–113 (2004)
4. Dräger, A., Rodriguez, N., Dumousseau, M., Dörr, A., Wrzodek, C., Le Novère, N., Zell, A., Hucka, M.: JSBML: a flexible Java library for working with SBML. Bioinformatics 27(15), 2167–2168 (2011)
5. Förster, J., Gombert, A.K., Nielsen, J.: A functional genomics approach using metabolomics and in silico pathway analysis. Biotechnology and Bioengineering 79(7), 703–712 (2002)
6. Funahashi, A.: CellDesigner2. 0: A Process Diagram Editor for Gene-regulatory and Biochemical Networks, p. 2–3
7. Glez-Peña, D., Reboiro-Jato, M., Maia, P., Rocha, M., Díaz, F., Fdez-Riverola, F.: AIBench: a rapid application development framework for translational research in biomedicine. Computer Methods and Programs in Biomedicine 98(2), 191–203 (2010)
8. Hucka, M., Finney, A., Sauro, H.M., Bolouri, H., Doyle, J.C., et al.: The systems biology markup language (SBML): a medium for representation and exchange of biochemical network models. Bioinformatics 19(4), 524–531 (2003)
9. Kanehisa, M., Goto, S.: KEGG: kyoto encyclopedia of genes and genomes. Nucleic Acids Research 28(1), 27–30 (2000)
10. Pavlopoulos, G.A., Wegener, A.-L., Schneider, R.: A survey of visualization tools for biological network analysis. BioData Mining 1, 12 (2008)
11. Raman, K., Chandra, N.: Flux balance analysis of biological systems: applications and challenges. Briefings in Bioinformatics 10(4), 435–449 (2009)
12. Rocha, I., Maia, P., Evangelista, P., Vilaça, P., Soares, S., Pinto, J.P., Nielsen, J., Patil, K.R., Ferreira, E.C., Rocha, M.: OptFlux: an open-source software platform for in silico metabolic engineering. BMC Systems Biology 4, 45 (2010)

13. Rocha, M., Maia, P., Mendes, R., Pinto, J.P., Ferreira, E.C., Nielsen, J., Patil, K.R., Rocha, I.: Natural computation meta-heuristics for the in silico optimization of microbial strains. BMC Bioinformatics 9, 499 (2008)
14. Shannon, P., Markiel, A., Ozier, O., Baliga, N.S., Wang, J.T., Ramage, D., Amin, N., Schwikowski, B., Ideker, T.: Cytoscape: a software environment for integrated models of biomolecular interaction networks. Genome Research 13(11), 2498–2504 (2003)
15. Stelling, J., Klamt, S., Bettenbrock, K., Schuster, S., Gilles, E.D.: Metabolic network structure determines key aspects of functionality and regulation. Nature 420(6912), 190–193 (2002)

A Workflow for the Application of Biclustering to Mass Spectrometry Data

Hugo López-Fernández[1], Miguel Reboiro-Jato[1], Sara C. Madeira[2],
Rubén López-Cortés[3], J.D. Nunes-Miranda[3], H.M. Santos[3],
Florentino Fdez-Riverola[1], and Daniel Glez-Peña[1]

[1] ESEI: Escuela Superior de Ingeniería Informática, University of Vigo,
Edificio Politécnico, Campus Universitario As Lagoas s/n, 32004, Ourense, Spain
{hlfernandez,mrjato,dgpena,riverola}@uvigo.es
[2] Knowledge Discovery and Bioinformatics group (KDBIO), INESC-ID
Instituto Superior Técnico (IST), Technical University of Lisbon, Lisbon, Portugal
sara.madeira@ist.utl.pt
[3] Bioscope Group, REQUIMTE, Departamento de Química,
Faculdade de Ciencias e Tecnologia (FCT), Universidade Nova de Lisboa
2829-516 Caparica, Portugal
{j.dinis.miranda,rlcortes}@uvigo.es, hms14862@fct.unl.pt

Abstract. Biclustering techniques have been successfully applied to analyze microarray data and they begin to be applied to the analysis of mass spectrometry data, a high-throughput technology for proteomic data analysis which has been an active research area during the last years. In this work, we propose a novel workflow to the application of biclustering to MALDI-TOF mass spectrometry data, supported by a software desktop application which covering all of its stages. We evaluate the adequacy of applying biclustering to analyze mass spectrometry by comparing between biclustering and hierarchical clustering over two real datasets. Results are promising since they revealed the ability of these techniques to extract useful information, opening a door to further works.

Keywords: biclustering, mass spectrometry, BiMS, BiBit, Bimax.

1 Introduction

In the last years, high-throughput mass spectrometry (MS) based proteomic data analysis has been an active research area. MS technology allows researchers to measure the mixture of peptides or proteins present in biological samples, such as urine or tissues. These measurements can be further used for discovering condition related patterns (biomarker discovery) [1] or sample classification [2], after a proper preprocessing of the raw data.

In the literature, different machine learning methods have been applied to biomarker discovery from mass spectrometry data [3]. Common approaches consist in employing unsupervised classification techniques such as hierarchical clustering (HC) [4] to extract natural clusters from the data, and then decide whether these correspond to the groups of study [5]. Given a preprocessed MS dataset, a clustering analysis will

M.S. Mohamad et al. (Eds.): *7th International Conference on PACBB*, AISC 222, pp. 145–153.
DOI: 10.1007/978-3-319-00578-2_19 © Springer International Publishing Switzerland 2013

group samples into clusters basing on all the detected peaks (masses). This grouping usually corresponds to majoritary classes or partitions in the dataset. However, it may be interesting to discover groups of samples with patterns only under specific masses, as they may lead to discover new sample groups associated by a previously unknown condition.

Biclustering techniques are appropriate to extract these local patterns (biclusters). This family of techniques has been successfully applied to analyze microarray data due to their ability to discover co-expressed genes under certain samples. In contrast to traditional clustering techniques, where each gene in a given cluster is defined under all the samples, biclustering algorithms propose groups of genes that show similar activity patterns under a subset of the experimental samples. Although it use is still minority, biclutering has recently started to be used in MS studies [6].

In this work we study the application of biclustering techniques to MS data, which can be useful to discover hidden phenotypes, in the form of relevant biclusters, for finding alternative hypothesis supported by the input data to classify available samples and subsequently, to identify potential biomarkers. This approach can be especially advantageous when dealing with unclassified data, or when the known classification/labeling cannot be achieved with traditional computational analysis. Although dimensionality in microarray data is higher (thousands of genes) compared to MS data (hundreds of peaks), we believe that MS data analysis also can take advantage of biclustering in discovering hidden hypotheses that classic partitioning approaches are not able to.

In this work, we propose a workflow to aid in the application of biclustering to mass spectrometry datasets. Moreover, MALDI-TOF MS datasets are analyzed following the proposed workflow using BiMS tool, an application specifically created to support this work.

The paper is structured as follows. Section 2 describes the proposed approach. Section 3 reviews the results. Finally, Section 4 concludes the paper and outlines future research work.

2   Applying Biclustering to MS Data

2.1   Proposed Workflow

The proposed workflow for biclustering construction using MS data (Fig. 1) is divided in two well differentiated steps: (*i*) preprocessing and (*ii*) biclustering analysis.

The preprocessing stage, which comprises several tasks that will be further explained in Subsection 2.2, aligns the masses of a given raw MS dataset. A typical MS dataset consists of a set of samples. Usually, for each sample, there are several *biological* replicates and, for each biological replicate, there are several *technical* replicates (i.e. the result of applying the same experimental procedure several times to the same biological replicate). Finally, each technical replication is spotted several times into the mass spectrometer. The first four preprocessing steps (intensity transformation, baseline correction, smoothing and peak detection) are applied to each spectrum individually, whereas intra-sample alignment is applied to the spot spectra of a technical replication, aligning it and reducing it into a single spectrum. Finally, all the samples are globally aligned (inter-sample alignment) so that peaks corresponding to the same compound are grouped together.

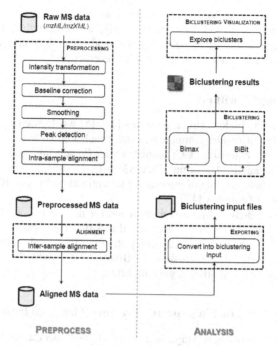

Fig. 1. Proposed workflow, divided into preprocess and analysis

Before the application of biclustering algorithms, the aligned MS data needs to be converted into a suitable format, as it will be explained in Subsection 2.4. Once the data has been converted, the selected biclustering algorithms are applied (see Subsection 2.3). Finally, the resulting biclusters are showed.

2.2   Preprocessing Mass Spectrometry Data

Preprocessing of MS data is a critical stage that transforms raw data into a suitable input for further analysis, such as machine learning or biomarker discovery. Inadequate or incorrect preprocessing methods can result in biased dataset and hinder to reach meaningful biological conclusions [7]. In this situation, preprocessing is necessary since raw data contains signals coming from the real peptides/proteins, as well as signals derived from several forms of noise (e.g. chemical, electronic factors, etc). The specific goals of this phase are (*i*) to remove noisy peaks without discarding any of the true peaks, and (*ii*) to determine the m/z and intensity values with the best accuracy [8].

Since there is no standard mass spectrometry data preprocessing pipeline, some authors proposed different guidelines to establish a design/data analysis protocol (DAP) [9, 10]. Based on these works, we decided to include the following preprocessing steps: (*i*) intensity transformation, (*ii*) baseline correction, (*iii*) smoothing, (*iv*) peak detection and (*v*) peak alignment.

In the last years, several algorithms and tools have been proposed to address each preprocessing task [11]. In this work, we opted to use MALDIquant [12], a versatile R package for the analysis of mass spectrometry data which covers the five preprocessing steps of the proposed workflow. Additionally, we have included the MassSpec-Wavelet R package [13], which performs the peak detection step without explicit intensity transformation, smoothing and baseline correction.

2.3   Biclustering Algorithms

Among all the existing biclustering algorithms [14, 15] we decided to use Bimax [16], which achieves similar results compared to other well known biclustering algorithms (Cheng and Church's algorithm, CC; Samba; Order Preserving Submatrix Algorithm, OPSM; Iterative Signature Algorithm, ISA; xMotif). An advantage of Bimax is that it is a powerful approach, capable of generating all optimal biclusters. Bimax is publicly available in the BicAT toolbox [17].

We also included BiBit [18] a novel approach for the extraction of biclusters from binary datasets. Domingo S. *et. al.* showed that BiBit can obtain similar results to Bimax by using significantly less computation time and reducing the total number of generated biclusters. BiBit is publicly available at http://www.upo.es/eps/bigs/BiBit_algorithm.html.

2.4   Searching for Mass Fingerprints (Presence/Abscense) Patterns

The result of the preprocessing stage is a set of aligned spectra, each one containing the identified peaks (m/z) and their corresponding intensity values, that must be converted into a suitable input for the biclustering algorithms.

The selected biclustering algorithms (Bimax and BiBit) take as input a binary dataset where 1 represents a peak presence, while 0 represents a peak absence. These algorithms will look for groups (biclusters) of 1's, that we call *presence patterns*. A set of aligned spectra can be simply transformed into a matrix where samples and masses are the dimensions (rows and columns) and each cell contains a 1 if the corresponding sample has the corresponding mass, and 0 otherwise.

In certain cases, it can be desirable to extract other type of patterns, such as *absence patterns* (biclusters of 0's) or *simple presence/absence patterns* (biclusters of 1's and 0's in one direction). Fig. 2 shows examples of the mentioned types of patterns.

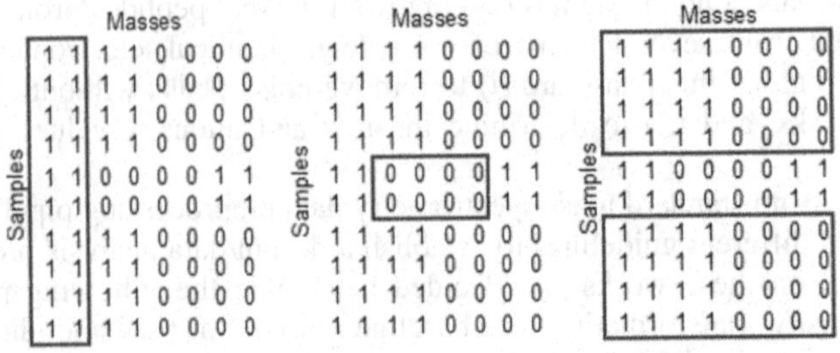

Fig. 2. From left to right: *presence*, *absence*, and *simple absence/presence* pattern

It is possible to extract this kind of patterns by only preprocessing the input binary matrix. By feeding the biclustering algorithm with the inverted data matrix, it will find *absence patterns*. Additionally, by joining the input data matrix and the inverted matrix, enables the extraction of *simple presence/absence patterns*.

3   Results and Discussion

In order to evaluate the adequacy of applying biclustering to analyze MS data, two datasets generated with MALDI-TOF mass spectrometer and taken from previous works [19, 20] were analyzed by using the proposed workflow. These works were selected to carry out the following analyses as they present two labeled datasets, that is, each sample belongs to a known biological condition or class. The former work [19] includes a smaller and simpler dataset with 12 samples and 3 different classes, while the latter [20] includes a larger dataset with 70 samples and 14 classes.

Additionally, a hierarchical clustering (HC) was also applied to evaluate the potential benefits of using biclustering techniques instead of other classical one-dimensional clustering techniques.

We will focus in finding *class-clusters*, that is, clusters including all the samples of a given class. Knowing the class associated with each sample will allow us to evaluate and compare each clustering and biclustering technique through the homogeneity of the formed clusters. In this work, we are especially interested in those classes where HC is unable to separate well.

The following subsections present the results of the described analyses. The datasets preprocessing and analysis were performed using the parameter configuration described in their respective published works, where possible.

3.1   Case Study I: Cancer Dataset

In the R. López-Cortés *et. al.* work [19], authors propose the use of gold-nanoparticles to separate the proteins and peptides in human serum as a way to improve MALDI-based sample profiling. The protocol described in this work divides each sample into two sub-samples: pellet and supernatant. Authors demonstrate that the spectra of both sub-samples generated by the MALDI can be grouped by their corresponding conditions using three-dimensional Principal Component Analysis (PCA).

In this study, the dataset is composed of 5 lymphoma samples, 5 myeloma samples and 2 healthy samples. As the aforementioned work shows that results using the pellet or supernatant sub-samples are similar, we decided to use only the latter sub-samples.

Both HC and biclustering were performed using BiMS. To perform the HC, first the dataset is loaded excluding those peaks with peak intensity lower than 0.175 and a signal-to-noise ratio (SNR) lower than 9. Then, HC analysis was done according the following parameters: (*i*) hamming distance, (*ii*) peptide mass tolerance of 150 ppm and (*iii*) a minimum intra-sample percentage of presence (POP) of 80%. As Fig. 3 shows, the three classes are well separated trough HC.

On the other hand, to carry out the biclustering analysis the dataset is loaded as described before. As the biclustering algorithms need the dataset to be pre-aligned, inter-sample alignment was done using MALDIquant with a tolerance of 0.002.

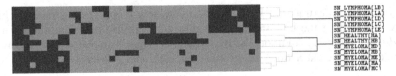

Fig. 3. Hierarchical clustering of the cancer dataset

Finally, applying the BiBit algorithm to find presence *class-clusters* with the default parameters, we are able to retrieve one bicluster containing the two healthy samples, one bicluster containing all the lymphoma samples, and another one containing 4 of the 5 myeloma samples (Fig. 4). Using Bimax the same results are achieved.

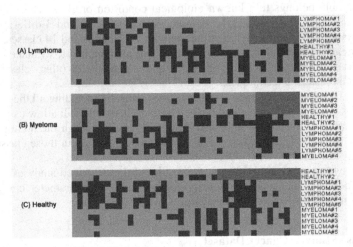

Fig. 4. Class-biclusters of the cancer dataset

3.2   Case Study II: Wine Dataset

In the J.D. Miranda Nunes *et. al.* work [20] authors propose the use of MALDI as a fast way to identify different wines types. As a part of this study, authors also evaluated the use of different matrix types and number of replicates per sample, determining that -Cyano is most suitable matrix and that better results are obtained with 5 replicates. Therefore, we have employed the dataset with these characteristics. This dataset is more complex, since it contains 5 samples of 14 different Spanish wines (70 samples in total).

Here we took the raw data and applied the following preprocessing using MALDI-quant: (*i*) square root intensity transformation, (*ii*) baseline correction in automatic mode, (*iii*), moving average as smoothing method, (*iv*) peak detection with SNR set to 3 and the half window size parameter set to 60, (*v*) intra-sample alignment with a tolerance of 0.002 and POP of 80%, and (*vi*) inter-sample alignment with a tolerance of 0.002.

Then, biclustering was done using Bimax according to the following configuration: (*i*) minimum number of samples in a bicluster was set to 4, (*ii*) minimum number of peaks in a bicluster was set to 2, (*iii*) "samples" biclustering mode, (*iv*) "presence" bicluster type and (*v*) search for *class-biclusters*.

Through this biclustering analysis, this preprocessed dataset allowed us to find *class-biclusters* for all the classes. There exist one bicluster containing all the samples in classes *A, D, E, F, G, J, K,* and *L,* and also several biclusters containing 4 of 5 samples for the remaining classes (i.e. *B, C, H, I, M,* and *N*). However, Bibit is able to extract *class-biclusters* only for classes *B, E, G, L,* and *K* with all the samples, and for classes *A, F,* and *I* with 4 samples.

In this case, the HC was also done using the same configuration as for the previous case study. This HC generated perfect groups for *B, C, G, I,* and *H* classes, whereas for classes *A, K, M,* and *E* groups with one missing sample were constructed. The remaining samples (i.e. *D, F, J, L,* and *N*) are mixed and cannot be grouped according to their labeling.

4   Conclusions

In this paper we presented a novel workflow for the application of biclustering to MS data, covering the preprocessing and the transformation of the data needed to perform the biclustering algorithms. Moreover, this work is supported by BiMS (http://sing.ei.uvigo.es/bims), an intuitive AIBench-based [21] desktop application which allows the user to manage and preprocess MS data, execute biclustering or HC algorithms, and explore the results.

The proposed workflow was evaluated using two MS datasets taken from previous studies and compared with HC. Although first dataset was preprocessed as in the original study and second dataset was preprocessed using the proposed workflow, in both cases biclustering techniques were able to identify *class-biclusters* for most of the classes.

Two different biclustering algorithms were tested: Bimax and BiBit. Both algorithms achieved similar results with the first and simpler dataset tested, while Bimax showed a better performance with the second dataset. Therefore, Bimax seems to be more powerful than BiBit, with the counterpart of needing more processing time.

Results also showed that, in certain cases, biclustering algorithms can extract more information than one-dimensional clustering techniques, although the generated biclusters must be filtered to select those of interest. As Bimax generates notably more biclusters than BiBit, this filtering step is easier with the latter algorithm.

Further work will focus on evaluate the quality of the resulting biclusters as a way to identify the most promising ones. This is especially interesting for non-labeled datasets, where biclusters can be taken as new hypothesis that may lead to discover new conditions or relationships. Additionally, we expect that using more complex presence/absence patterns, as described in Subsection 2.4, will allow us to discover more valuable relations that simple presence patterns cannot found.

Finally, we can conclude that biclustering is a promising technique in MS, able to unveil hidden hypotheses in the form of sample groups whose similarity patterns are local across the features (i.e. peaks), instead of global. Moreover, since proteins usually express more than one peak, a per-protein bicluster could be discovered. Therefore, given the overlapping nature of biclustering, a sample could belong to more than one bicluster, because it expresses more than one protein, which is a biologically interpretable hypothesis.

Acknowledgements. This work is partially funded by the (*i*) TIN2009-14057-C03-02 project from the Spanish Ministry of Science and Innovation, the Plan E from the Spanish Government and the European Union from the ERDF, (*ii*) the integrated action AIB2010PT-00353 from the Spanish Ministry of Science and Innovation, (*iii*) Agrupamento INBIOMED (2012/273) from DXPCTSUG-FEDER unha maneira de facer Europa and (*iv*) Portuguese funds through FCT – Fundação para a Ciência e a Tecnologia, under projects PEst-OE/EEI/LA0021/2011 and PTDC/EIA-EIA/111239/2009. H. López-Fernández was supported by a pre-doctoral fellowship from the University of Vigo.

References

1. Roy, P., Truntzer, C., Maucort-Boulch, D., Jouve, T., Molinari, N.: Protein mass spectra data analysis for clinical biomarker discovery: A global review. Briefings Bioinf. 12(2), 176–186 (2011)
2. Tibshirani, R., Hastie, T., Narasimhan, B., Soltys, S., Shi, G., Koong, A., Le, Q.T.: Sample classification from protein mass spectrometry, by 'peak probability contrasts". Bioinformatics 20(17), 3034–3044 (2004)
3. Diamandis, E.: Mass spectrometry as a diagnostic and a cancer biomarker discovery tool: Opportunities and potential limitations. Expert Syst. Appl. 3(4), 367–378 (2004)
4. Yang, P., Zhang, Z., Zhou, B.B., Zomaya, A.Y.: A clustering based hybrid system for biomarker selection and sample classification of mass spectrometry data. Neurocomputing 73(13-15), 2317–2331 (2010)
5. McDonald, R., Skipp, P., Bennell, J., Potts, C., Thomas, L., O'Connor, C.D.: Mining whole-sample mass spectrometry proteomics data for biomarkers – An overview. Expert Syst. Appl. 36(3), 5333–5340 (2009)
6. Choi, H., Kim, S., Gingras, A.C., Nesvizhskii, A.: Analysis of protein complexes through model-based biclustering of label-free quantitative AP-MS data. Mol. Syst. Biol. 6, 385 (2010)
7. Coombes, K.R., Baggerlyand, K.A., Morris, J.S.: Pre-Processing Mass Spectrometry Data. In: Dubitzky, M., Granzow, M., Berrar, D. (eds.) Fundamentals of Data Mining in Genomics and Proteomics. Kluwer, Boston (2007)
8. Eidhammer, I., Flikka, K., Martens, L., Mikalsen, S.: Computational Methods for Mass Spectrometry Proteomics. Jon Wiley & Sons, Ltd., England (2008)
9. Armananzas, R., Saeys, Y., Inza, I., Garcia-Torres, M., Bielza, C., van de Peer, Y., Larranaga, P.: Peakbin selection in mass spectrometry data using a consensus approach with estimation of distribution algorithms. IEEE/ACM Trans. Comput. Biol. Bioinf. 8(3), 760–774 (2011)
10. Barla, A., Jurman, G., Riccadonna, S., Merler, S., Chierici, M., Furlanello, C.: Machine learning methods for predictive proteomics. Briefings Bioinf. 9(2), 119–128 (2008)
11. Yang, C., He, Z., Yu, W.: Comparison of public peak detection algorithms for MALDI mass spectrometry data analysis. BMC Bioinf. 10, 4 (2009)

12. Du, P., Kibbe, W.A., Lin, S.M.: Improved peak detection in mass spectrum by incorporating continuous wavelet transform-based pattern matching. Bioinformatics 22(17), 2059–2065 (2006)
13. Gibb, S., Strimmer, K.: MALDIquant: a versatile R package for the analysis of mass spectrometry data. Bioinformatics 28(17), 2270–2271 (2012)
14. Madeira, S.C., Oliveira, A.L.: Biclustering Algorithms for Biological Data Analysis: A Survey. IEEE/ACM Trans. Comput. Biol. Bioinf. I(I), 24–45 (2004)
15. Verma, N.K., Meena, S., Bajpai, S., Singh, A., Nagrare, A., Cui, Y.: A Comparison of Biclus-tering Algorithms. In: Proceedings of the Int. Conf. Syst. Med. Biol. (ICSMB 2010), pp. 90–97 (2010)
16. Prelić, A., Bleuler, S., Zimmermann, P., Wille, A., Bühlmann, P., Gruissem, W., Hennig, L., Thiele, L., Zitzler, E.: A systematic comparison and evaluation of biclustering methods for gene expression data. Bioinformatics 22(9), 1122–1129 (2006)
17. Barkow, S., Bleuler, S., Prelic, A., Zimmermann, P., Zitzler, E.: BicAT: a biclustering analysis toolbox. Bioinformatics 22(10), 1282–1283 (2006)
18. Rodriguez-Baena, D.S., Perez-Pulido, A.J., Aguilar-Ruiz, J.S.: A biclustering algorithm for extracting bit-patterns from binary datasets. Bioinformatics 27(19), 2738–2745 (2001)
19. López-Cortés, R., Oliveira, E., Núñez, C., Lodeiro, C., Páez de la Cadena, M., Fdez-Riverola, F., López-Fernández, H., Reboiro-Jato, M., Glez-Peña, D., Capelo, J.L., Santos, H.M.: Fast human serum profiling through chemical depletion coupled to gold-nanoparticle-assisted protein separation. Talanta 100, 239–245 (2012)
20. Nunes-Miranda, J.D., Santos, H.M., Reboiro-Jato, M., Fdez-Riverola, F., Igrejas, G., Lodeiro, C., Capelo, J.L.: Direct matrix assisted laser desorption ionization mass spectrometry-based analysis of wine as a powerful tool for classification purposes. Talanta 91, 72–76 (2012)
21. Glez-Peña, D., Reboiro-Jato, M., Maia, P., Díaz, F., Fdez-Riverola, F.: AIBench: a rapid application development framework for translational research in biomedicine. Comput. Meth. Prog. Bio. 98, 191–203 (2010)

Author Index